T0261382

Advanced Ceramic Coatings and Materials for Extreme Environments III

Advanced Ceramic Coatings and Materials for Extreme Environments III

A Collection of Papers Presented at the 37th International Conference on Advanced Ceramics and Composites January 27–February 1, 2013 Daytona Beach, Florida

Edited by
Hua-Tay Lin
Taejin Hwang

Volume Editors
Soshu Kirihara
Sujanto Widjaja

The American Ceramic Society

WILEY

Cover Design: Wiley

Copyright © 2014 by The American Ceramic Society. All rights reserved.

Published by John Wiley & Sons, Inc., Hoboken, New Jersey.
Published simultaneously in Canada.

No part of this publication may be reproduced, stored in a retrieval system, or transmitted in any form or by any means, electronic, mechanical, photocopying, recording, scanning, or otherwise, except as permitted under Section 107 or 108 of the 1976 United States Copyright Act, without either the prior written permission of the Publisher, or authorization through payment of the appropriate per-copy fee to the Copyright Clearance Center, Inc., 222 Rosewood Drive, Danvers, MA 01923, (978) 750-8400, fax (978) 750-4470, or on the web at www.copyright.com. Requests to the Publisher for permission should be addressed to the Permissions Department, John Wiley & Sons, Inc., 111 River Street, Hoboken, NJ 07030, (201) 748-6011, fax (201) 748-6008, or online at http://www.wiley.com/go/permission.

Limit of Liability/Disclaimer of Warranty: While the publisher and author have used their best efforts in preparing this book, they make no representations or warranties with respect to the accuracy or completeness of the contents of this book and specifically disclaim any implied warranties of merchantability or fitness for a particular purpose. No warranty may be created or extended by sales representatives or written sales materials. The advice and strategies contained herein may not be suitable for your situation. You should consult with a professional where appropriate. Neither the publisher nor author shall be liable for any loss of profit or any other commercial damages, including but not limited to special, incidental, consequential, or other damages.

For general information on our other products and services or for technical support, please contact our Customer Care Department within the United States at (800) 762-2974, outside the United States at (317) 572-3993 or fax (317) 572-4002.

Wiley also publishes its books in a variety of electronic formats. Some content that appears in print may not be available in electronic formats. For more information about Wiley products, visit our web site at www.wiley.com.

Library of Congress Cataloging-in-Publication Data is available.

ISBN: 978-1-118-80755-2
ISSN: 0196-6219

Printed in the United States of America.

10 9 8 7 6 5 4 3 2 1

Contents

Preface

This proceedings issue contains contributions from primarily two advanced coating related symposia, Next Advanced Ceramic Coatings for Structural, Environmental & Functional Applications and Generation Technologies for Innovative Surface Coatings that were part of the 37th International Conference on Advanced Ceramics and Composites (ICACC), in Daytona Beach, Florida, January 27-February 1, 2013. These symposia were sponsored by the ACerS Engineering Ceramics Division. These papers provide the up-to-date summary on the development and applications of engineering and functional ceramic coatings.

We are greatly in debt to the members of the symposium organizing committees, for their assistance in developing and organizing these vibrant and cutting-edge symposia. We also would like to express our sincere thanks to manuscript authors and reviewers, all the symposium participants and session chairs for their contributions to a successful meeting. Finally, we are also very grateful to the staff of The American Ceramic Society for their dedicated efforts in ensuring an enjoyable as well as successful conference and the high-quality publication of the proceeding volume.

H. T. LIN, *Oak Ridge National Laboratory, USA*
TAEJIN HWANG, *Korea Institute of Industrial Technology, Korea*

Introduction

This issue of the Ceramic Engineering and Science Proceedings (CESP) is one of nine issues that has been published based on manuscripts submitted and approved for the proceedings of the 37th International Conference on Advanced Ceramics and Composites (ICACC), held January 27–February 1, 2013 in Daytona Beach, Florida. ICACC is the most prominent international meeting in the area of advanced structural, functional, and nanoscopic ceramics, composites, and other emerging ceramic materials and technologies. This prestigious conference has been organized by The American Ceramic Society's (ACerS) Engineering Ceramics Division (ECD) since 1977.

The 37th ICACC hosted more than 1,000 attendees from 40 countries and approximately 800 presentations. The topics ranged from ceramic nanomaterials to structural reliability of ceramic components which demonstrated the linkage between materials science developments at the atomic level and macro level structural applications. Papers addressed material, model, and component development and investigated the interrelations between the processing, properties, and microstructure of ceramic materials.

The conference was organized into the following 19 symposia and sessions:

Symposium 1	Mechanical Behavior and Performance of Ceramics and Composites
Symposium 2	Advanced Ceramic Coatings for Structural, Environmental, and Functional Applications
Symposium 3	10th International Symposium on Solid Oxide Fuel Cells (SOFC): Materials, Science, and Technology
Symposium 4	Armor Ceramics
Symposium 5	Next Generation Bioceramics
Symposium 6	International Symposium on Ceramics for Electric Energy Generation, Storage, and Distribution
Symposium 7	7th International Symposium on Nanostructured Materials and Nanocomposites: Development and Applications

Symposium 8	7th International Symposium on Advanced Processing & Manufacturing Technologies for Structural & Multifunctional Materials and Systems (APMT)
Symposium 9	Porous Ceramics: Novel Developments and Applications
Symposium 10	Virtual Materials (Computational) Design and Ceramic Genome
Symposium 11	Next Generation Technologies for Innovative Surface Coatings
Symposium 12	Materials for Extreme Environments: Ultrahigh Temperature Ceramics (UHTCs) and Nanolaminated Ternary Carbides and Nitrides (MAX Phases)
Symposium 13	Advanced Ceramics and Composites for Sustainable Nuclear Energy and Fusion Energy
Focused Session 1	Geopolymers and Chemically Bonded Ceramics
Focused Session 2	Thermal Management Materials and Technologies
Focused Session 3	Nanomaterials for Sensing Applications
Focused Session 4	Advanced Ceramic Materials and Processing for Photonics and Energy
Special Session	Engineering Ceramics Summit of the Americas
Special Session	2nd Global Young Investigators Forum

The proceedings papers from this conference are published in the below nine issues of the 2013 CESP; Volume 34, Issues 2–10:

- Mechanical Properties and Performance of Engineering Ceramics and Composites VIII, CESP Volume 34, Issue 2 (includes papers from Symposium 1)
- Advanced Ceramic Coatings and Materials for Extreme Environments III, Volume 34, Issue 3 (includes papers from Symposia 2 and 11)
- Advances in Solid Oxide Fuel Cells IX, CESP Volume 34, Issue 4 (includes papers from Symposium 3)
- Advances in Ceramic Armor IX, CESP Volume 34, Issue 5 (includes papers from Symposium 4)
- Advances in Bioceramics and Porous Ceramics VI, CESP Volume 34, Issue 6 (includes papers from Symposia 5 and 9)
- Nanostructured Materials and Nanotechnology VII, CESP Volume 34, Issue 7 (includes papers from Symposium 7 and FS3)
- Advanced Processing and Manufacturing Technologies for Structural and Multi functional Materials VII, CESP Volume 34, Issue 8 (includes papers from Symposium 8)
- Ceramic Materials for Energy Applications III, CESP Volume 34, Issue 9 (includes papers from Symposia 6, 13, and FS4)
- Developments in Strategic Materials and Computational Design IV, CESP Volume 34, Issue 10 (includes papers from Symposium 10 and 12 and from Focused Sessions 1 and 2)

The organization of the Daytona Beach meeting and the publication of these proceedings were possible thanks to the professional staff of ACerS and the tireless dedication of many ECD members. We would especially like to express our sincere thanks to the symposia organizers, session chairs, presenters and conference attendees, for their efforts and enthusiastic participation in the vibrant and cutting-edge conference.

ACerS and the ECD invite you to attend the 38th International Conference on Advanced Ceramics and Composites (http://www.ceramics.org/daytona2014) January 26-31, 2014 in Daytona Beach, Florida.

To purchase additional CESP issues as well as other ceramic publications, visit the ACerS-Wiley Publications home page at www.wiley.com/go/ceramics.

SOSHU KIRIHARA, *Osaka University, Japan*
SUJANTO WIDJAJA, *Corning Incorporated, USA*

Volume Editors
August 2013

PROGRESS IN EBC DEVELOPMENT FOR SILICON-BASED, NON-OXIDE CERAMICS

C.A. Lewinsohn, H. Anderson, J. Johnston
Ceramatec Inc.
Salt Lake City, UT, 84119

Dongming Zhu
NASA Glenn Research Center
Cleveland, OH

Hydrothermal corrosion is a lifetime-limiting mechanism for silicon-based, non-oxide ceramics in combustion environments. Many desirable materials for use as protective coatings are physically or chemically incompatible with the non-oxide substrate materials. A unique method of engineering bond-coats and coating systems for non-oxide systems has been developed and shown to improve the hydrothermal corrosion resistance of silicon nitride and silicon carbide-based materials. Progress in work to investigate the effect of additions of oxidation resistant filler materials to polymer-derived bond coats for environmental barrier coatings will be discussed. In the current work, additional data will be provided showing that the bond coat system can be adapted to composite silicon carbide. Initial results on the high-temperature durability of these coatings will be presented.

INTRODUCTION

Higher turbine inlet temperatures are one way of making turbine engines consume less fuel, which reduces both operating costs and emissions. Currently, however, turbine inlet temperatures are limited to values below which downstream components can survive. Typically, creep of metallic components at elevated temperatures and pressures limit the conditions at which turbines can be operated. Therefore, for many years, there has been a desire to introduce into turbine engine hot sections ceramic materials that are more resistant to creep. Even after sufficient strength, creep behavior, and reliability had been demonstrated, however, the lifetime of the candidate materials was found to be below desired values due to corrosion in the presence of water vapor[1], which is referred to as hydrothermal corrosion.

The ceramic materials that have sufficient strength, creep behavior, and reliability to withstand turbine engine conditions are typically silicon based ceramics, such as silicon nitride and silicon carbide, and their composites. Silicon carbide (SiC) has high strength and good thermal conductivity, but it suffers from low fracture toughness and, hence, reliability. Therefore, components consisting of silicon nitride (Si_3N_4), which can be manufactured with higher values of fracture toughness than silicon carbide, and silicon carbide- or silicon nitride–matrix composites are currently under development for components that will be subject to appreciable stresses in operation. These materials are stable under purely oxidizing conditions, due to the formation of passivating oxide layers. In processes known as hydrothermal corrosion, however, these materials be corroded significantly by H_2O and CO, which are common components in gas turbine systems[2,3,4].

Extensive research has identified several oxide materials with low silica-activity that are relatively resistant to hydrothermal corrosion. These materials do not possess the strength, creep behavior and reliability required to act as the structural component, however they could be used as coatings for turbine engine hot section components. These oxides include ytterbium silicate

($Yb_2Si_2O_7$)[8, 5, 6], lutetium silicate ($Lu_2Si_2O_7$)[6], yttria-stabilised zirconia (8mol% yttria + 92mol% ZrO_2, 8YSZ)[5], strontium-stabilised celsian (($1-x$)BaO-xSrO-AlO2-SiO2, $0<x<1$), BSAS)[9], and mullite ($3Al_2O_3$-$2SiO_2$).[7]

The oxide materials that exhibit good resistance to hydrothermal corrosion typically have much higher coefficients of thermal expansion than silicon nitride[8] or silicon-based materials, such that unacceptably high residual stresses develop in the substrate or coating that subsequently lead to failure after processing or during operation. One approach to mitigate these residual stresses has been to insert materials with intermediate properties between the coating and the substrate.[9] The interlayer material has been limited by the requirements that it adheres to both top coat and substrate materials, has good high-temperature stability, does not exhibit any deleterious reactions with either the top coat or substrate, and has acceptable thermoelastic properties. Over the past decade, development of amorphous, non-oxide ceramics derived from preceramic polymers (polymer-derived ceramics, PDC) has led to materials with remarkable oxidation stability and mechanical properties at elevated temperatures.[10] Furthermore, these materials show excellent adherence to a wide range of materials, including non-oxide ceramics, oxide ceramics, and metals. Therefore, efforts at Ceramatec, Inc. have been focused on developing coating systems with PDC interlayers and low silica-activity, outer environmental barrier coatings (EBC). This paper will describe recent progress in the development of these coating systems and their application to silicon-based ceramics, including silicon carbide, fiber-reinforced, silicon carbide matrix composites (SiC_f/SiC_m).

METHODS

A simplified flow chart illustrating the overall process to coat substrates, is shown in Figure 1. This flow sheet illustrates the process used to apply bond coat layers to samples at Ceramatec, some of which were then sent to the NASA Glenn Research Center for application of the EBC layer. In earlier work, Ceramatec has also developed a method of cosintering bond coat layers with EBC layers. The latter method may be more desirable since the issue of sintering material onto a non-densifying substrate only occurs once during processing. Regardless, it is desirable to have a low modulus layer in the coating system to accommodate property mismatches, so ideal coating systems have dense, outer EBC layers and bond coat layers with residual porosity.

Surface preparation is dependent on the nature of the surface of the substrate being coating, but at a minimum it involves cleaning and degreasing the surfaces to be coated and at a maximum involves sandblasting the surface to create a new one. The bond coat can be applied as several layers, with the same or different composition, or as a single layer. At Ceramatec, efforts have focused on dip-coating the coating layers. Although dip coating has the advantage that it is not a line of sight process and is relatively simple, more sophisticated methods, such as spray deposition, may produce more uniform coatings or may have other benefits in manufacturing.

Coatings were applied to bend bars made from various commercial types of silicon nitride or silicon carbide, fiber-reinforced, silicon carbide matrix composites. In some cases the silicon nitride bend bars had as-processed surfaces, but in most cases the surfaces were machined and ground. The surfaces of the SiC_f/SiC_m bend bars were cleaned prior to coating, but no other treatment was applied to them.

Bond coat materials consisted predominantly of allyl-hydrido-polycarbosilane (aHPCS, Starfire Corp.), a preceramic polymer. A portion of the preceramic polymer was partially pyrolysed to convert it from liquid form to solid. The partially pyrolysed material was blended

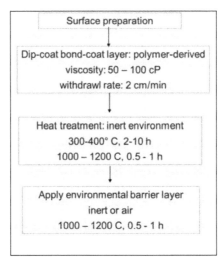

Figure 1   Simplified process flow chart for applying bond coat(s) and outer coatings, as required.

with liquid precursor and filler materials. Initial work at Ceramatec to produce co-fired coating systems utilized silicon nitride and EBC powder as filler material. For bond coats applied to silicon nitride at Ceramatec and sent to NASA for application of the EBC, silicon nitride powder and EBC powder was also used as fillers, however, removing the EBC powder from the bond coat provided the best results in high temperature thermal cycling. For SiC_f/SiC_m, five different blends of filler powder were evaluated separately: partially-pyrolysed preceramic precursor and silicon carbide, HfO_2 and silicon carbide, $Yb_2Si_2O_7$ and silicon carbide, HfO_2-Y_2O_3-GdO-$Yb_2Si_2O_7$ and silicon carbide, and silicon carbide nanotubes. The ratio of experimental filler/silicon carbide filler was kept constant, as was the volume fraction of the experimental filler. Non-aqueous mixtures, using toluene as a solvent, of the liquid preceramic precursor, partially pyrolysed precursor, and filler materials were prepared such that the viscosity of the mixtures was in the range of 50 – 100 cP. Samples were attached to a labscale dip-coating apparatus that immersed and withdrew the samples from the mixture at a constant rate, 2 cm/min. The samples were approximately 4 mm-wide, 3 mm-thick, and 50 mm-long.

The flexural strength of the bars was tested according to ASTM C-1160. Hydrothermal corrosion testing was performed in an environment of 90% H_2O, 10% O_2 flowing at 2.2 cm/sec. Thermal cycles were performed by shuttling the specimens in and out of the hot zone of a furnace held at the test temperature, i.e. 1300°C or 1382°C. The specimens were cycled between room temperature and the test temperature with a heating time of 20 seconds to temperature, a 1 h hold, a cooling time of several minutes, and a 20 minute hold at room temperature. The specimens were contained in platinum crucibles that were contained within an alumina furnace tube. The testing was performed at the NASA Glenn Research Center.

RESULTS AND DISCUSSION

Bond coats were applied to silicon nitride by dip-coating, Figure 2. The original bond coat formulation was derived for silicon nitride made by Saint-Gobain, Inc. (now owned by CoorsTek) and is referred to as "NT-154." When the same formulation was applied to another vendor's silicon nitride, extensive cracking occurred. The second silicon nitride was made by Kyocera, Inc. and is referred to as "SN-282". Modifications were made to the bond coat formulation to increase the porosity of the bond coat layer in a controlled manner so that it would be more tolerant to the differences in thermomechanical properties of the two substrate materials. These modified bond coats are referred to as "PDBC1" and "PDBC2". Flexural strength testing at NASA Lewis Research Center showed that applying their standard EBC system, referred to as "EPM", a pure-mullite EBC, or a silicon bond coat to SN282 greatly reduced its strength, Figure 3a. On the other hand, applying the PDC bond coat, or the PDC bond coat and EBC coatings did not affect the strength of the substrate material, Figure 3b.

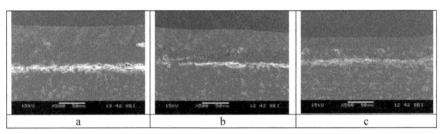

Figure 2   Scanning electron micrographs of polymer-derived bond coats (a) derived for NT-154 , (b) modification 1 for SN282, and (c) modification 2 for SN282.

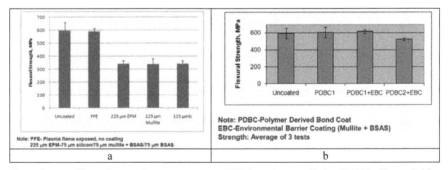

Figure 3   Flexural strength measurements of EBC coatings applied to SN282 silicon nitride (a) without and (b) with a PDC bond coat.

NASA demonstrated that both their standard EBC system, EPM, and rare-earth disilicate EBC material (REDS) could be deposited onto silicon nitride substrates if a PDC bond coat was applied, Figure 4 and Figure 5. Testing at NASA also showed that the EBC coatings, deposited on silicon nitride with the PDC bond coat, offered environmental resistance during thermal cycle

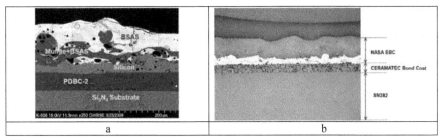

Figure 4   Images of NASA's EPM EBC applied to silicon nitride with a PDC bond coat before, (a), and after, (b), thermal cycle testing between room temperature and 1300°C.

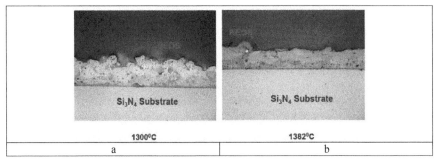

Figure 5   Images of NASA's EPM EBC applied to silicon nitride with a PDC bond coat after thermal cycling to 1300°C, (a), and 1382°C, (b).

testing up to 1300°C, Figure 5a. At 1382°C, however, the REDS EBC showed signs of cracking when examined after flexural testing, Figure 5b.

Subsequent to testing PDC bond coats and EBC for silicon nitride, investigation of the efficacy of PDC bond coats for SiC_f/SiC_m was performed. These efforts were focused on enabling several potential EBC systems for SiC_f/SiC_m to be evaluated. Therefore, PDC bond coats with filler powder made from the desired EBC materials were applied to composite substrates. As shown in Figure 6, although the PDC material produced a continuous bond coat, the initial EBC material applied to it cracked during densification. These results indicate that additional adjustments, similar to those made when the bond coat formulation used initially for silicon nitride NT-154 was modified for silicon nitride SN282, are required for these candidate EBC. Regardless of the presence of cracks, the room temperature flexural strength of the composites was not degraded by the application of the coatings, Table 1.

Table 1

	Avg. Strength (MPa)	95% C.I.
HfO_2	196	27
Re-HfO_2	210	28

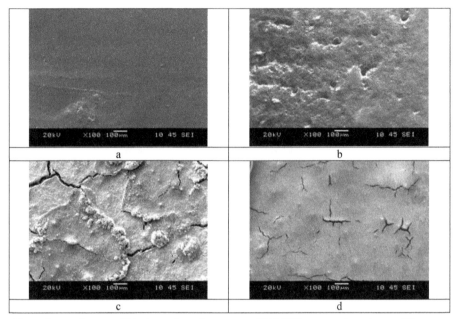

Figure 6   Images of (a), the surface of an uncoated SiC$_f$/SiC$_m$ composite substrate, (b), a SiC$_f$/SiC$_m$ substrate coated with PDC bond coat, (c) a SiC$_f$/SiC$_m$ sample coated with a PDC bond coat and HfO$_2$ EBC, and (d) a SiC$_f$/SiC$_m$ sample coated with a PDC bond coat and rare earth-doped, HfO$_2$ EBC.

SUMMARY

A method for making bond coat materials and EBC systems for protection from hydrothermal corrosion of silicon-based ceramics has been demonstrated. The method that has been developed allows for tailoring to specific substrate materials and specific EBC materials. Application of the coatings does not damage substrate materials. Coatings using the PDC bond coat have survived thermal cycling up to 1300°C. Additional development is required to verify performance with state of the art substrate and EBC materials.

ACKNOWLEDGEMENTS

The authors are grateful for support for this work from the NASA Lewis Research Center.

REFERENCES

[1]   N.S. Jacobson, "Corrosion of Silicon-Based Ceramics in Combustion Environments," *J. Am. Ceram. Soc.*, 76 [1] 3–28 (1993).

[2]   E.J. Opila and N. S. Jacobson, "SiO(g) formation from SiC in Mixed Oxidizing/reducing Gases," *Oxidation of Metals* **44** [5/6] 527-544 (1995).

[3]   E.J. Opila, J.L. Smialek, R.C. Robinson, D.S. Fox, N.S. Jacobson, "SiC Recession Caused by SiO2 Scale Volatility under Combustion Conditions: II, Thermodynamics and Gaseous-Diffusion Model," *J. Am. Ceram. Soc..*, **82** [7] 1826-1834 (1999).

[4]   E. J. Opila, "Oxidation and Volatilization of Silica Formers in Water Vapor", *J. Am. Ceram. Soc.,* 86 [8] 1238–48 (2003).

[5]   T. Fukudome, S. Tsuruzono, T. Tatsumi, Y. Ichikawa, T. Hisamatsu, I. Yuri, "Developments of Silicon Nitride Components for Gas Turbine," presented at ISASC-2004/International Symposium of New Frontier of Advanced Si-based Ceramics and Composites, Gyeongju, Kore, June 2004.

[6]   S. Ueno, N. Kondo, T. Ohji, S. Kanzaki, J. Doni, "High Temperature Hydro Corrosion Resistance of Silica Based Oxide Ceramics", American Society of Mechanical Engineers, International Gas Turbine Institute, Turbo Expo (Publication) IGTI, v 1, 2003, p 625-632.

[7]   K. N. Lee, "Key Durability Issues With Mullite-Based Environmental Barrier Coatings for Si-Based Ceramics", Journal of Engineering for Gas Turbines and Power, Vol. 122, 632-636, October (2000).

[8]   S. Ueno, D. Jayaseelan, N. Kondo, T. Ohji, and H.-T. Lin, "Development of EBC for Silicon Nitride," presented at ISASC-2004/International Symposium of New Frontier of Advanced Si-based Ceramics and Composites, Gyeongju, Kore, June 2004.

[9]   K. N. Lee, "Current Status of Environmental Barrier Coatings for Si-Based Ceramics," *Surf. Coatings Tech.* **133-134** 1-7 (2000).

[10]   R.Riedel, A. Kienzle, W. Dressler, L. Ruwisch, J. Bill, F. Aldinger, "Silicoboron carbonitride ceramic stable to 2,000 °C," *Nature*, **382**, 796 (1996).

FABRICATION OF SLURRY BASED Y-Si-Al-O ENVIRONMENTAL BARRIER COATING ON THE POROUS Si$_3$N$_4$ CERAMICS

Yinchao Liu, Chao Wang, Xuefen Lu, Hongjie Wang[*]
State Key Laboratory for Mechanical Behavior of Materials, Xi'an Jiaotong University, Xi'an, Shannxi 710049, China

ABSTRACT

The dense Y-Si-Al-O environment barrier coatings were fabricated on the surface of porous Si$_3$N$_4$ substrate by slurry spray method and then sintered at 1350℃-1450℃ in N2 atmosphere. Results show that different phase composition and microstructure of coatings were formed due to the different sintering temperature. The Y-Si-Al-O powder melted and infiltrated into the porous substrate and formed a dense layer with a thickness of about 100μm. The coating sintered at 1400℃ is uniform and crack-free with a composition of β-Si$_3$N$_4$ and part of the glass phase. After sintering, the water absorption decreased significantly from 34.5% to only 2.6% and the flexural strength was improved by 15.6%. The dielectric constant and loss of specimen increased due to the decrease of the porosity after coating.
KEY WORDS Environmental barrier coatings (EBC); Porous Si$_3$N$_4$ ceramics; Y-Si-Al-O coatings; Water absorption; Flexural strength; Dielectric constant and loss.

INTRODUCE

Porous Si$_3$N$_4$ ceramics is a promising broadband wave-transparent candidate material for the high temperature radome applications, owing to the good thermo-mechanical and dielectric properties.[1-5] Particularly, the low dielectric constant and loss ($\varepsilon'\approx$2.7-3.3, tanδ$\approx$ (5-10) ×10^{-4}, porosity≈40-55%) attracted the extensive research interest.[3,5,6] The porous structure is an easy way to obtain relatively low dielectric constant and loss,[6,7] but also is the cause of all the discordances. The water vapor can be easily collected from the moisture environment due to the characteristic of the porous structure. This is fatal to the dielectric properties of porous Si$_3$N$_4$ components. Furthermore, the porous structure will weaken the strength of the porous Si$_3$N$_4$ ceramics, making the porous Si$_3$N$_4$ radome components vulnerable for the erosion and corrosion of the rain drops or dusts. Therefore, it is crucial to fabricate a dense and uniformed environmental barrier coating (EBC) on the surface of porous Si$_3$N$_4$ ceramics.[1,5] To date, there have been few considerable developments about the EBCs on the porous Si$_3$N$_4$ substrate. Q Shen et al.[8] have sintered a layer of Si$_3$N$_4$ based coating by adding MgO, Al$_2$O$_3$, and SiO$_2$ as the sintering additives. XM Li et al.[5] have investigated deposition of Si$_3$N$_4$, BN, and B$_4$C coatings on the porous Si$_3$N$_4$ ceramic by CVD approach. And Chaochen Zhang et al.[9] fabricated a coating of β-SiAlON and O'-SiAlON by dip-coating with Sol-Gel slurry. All the reported results show that a cost-efficient coating which meets the desired requirements is in urgent need.

In this work, Y$_2$O$_3$, SiO$_2$ and Al$_2$O$_3$ were chosen as the initial powders in order to keep chemical consistency with matrix because the porous substrate was sintered with Y$_2$O$_3$/Al$_2$O$_3$ additives. To decrease the eutectic temperature of the system, SiO$_2$ was added into the mixture of Y$_2$O$_3$ and Al$_2$O$_3$ raw powder. The Y$_2$O$_3$-SiO$_2$-Al$_2$O$_3$ EBC system under certain ternary proportion

---

[*] Corresponding author. Tel: +86 (029) 82667942; email: hjwang@mail.xjtu.edu.cn

can be sintered at range of 1300℃ to 1500℃ according to the phase diagram.[10] The macro- and microstructure of the sintered Y-Si-Al-O coatings, as well as the related properties were measured to determine the qualities of the coatings.

EXPERIMENTAL

The Si_3N_4 substrate was prepared by Gel-casting and sintering at 1800℃ for 2h (β-Si_3N_4 phase, porosity≈53.0%, water absorption≈34.5%, flexural strength 35.2±2.5MPa). The substrate was cut and ground into bars and wafers with the dimensions of 3mm×4mm×40mm and Φ51mm×2.5mm respectively. The bars were used for bending test and the wafers for complex permittivity measurement.

The raw materials for the present work were Y_2O_3 (0.5μm，99.9% purity), amorphous SiO_2 (1.0μm, 99.0% purity), Al_2O_3 (1.0μm, 99.0% purity). The weight ratio of $Y_2O_3:SiO_2:Al_2O_3=$ 34.04:45.32:20.64. After ball milling with ethyl alcohol for 8h, the homogeneous mixture was precalcined in air at 1300℃ for 4h, and then crushed by ball milling with ethyl alcohol for 4-5 days to get the Y-Si-Al-O fine powder. Stable ethanol slurry with the solid content of 30 wt.% was ready for spraying after ball milled for 8h, and then deposited onto the substrate specimens by an air spray gun, whose parameters were given in the Table 1. After drying in the vacuum oven, the coating specimens were sintered in accordance with optimized temperature schedules at 1350℃, 1400℃, 1450℃ for 1h in 0.15MPa N_2 atmosphere. The heating and cooling rate between 1000℃and 1300℃ was 5℃/min, and 2℃/min when temperature was above 1300℃. To facility the description and discussion, the coatings were marked as C1350, C1400 and C1450 according to the different sintering temperature.

Table 1. Parameters of the air spray gun

Nozzle Diameter (mm)	Air Pressure (MPa)	Spraying Distance (cm)	Scanning Speed of Gun (cm/s)
0.5	0.4~0.7	20~25	10~20

The macro- and microstructure morphologies of the coating were detected by SEM (Quanta 600FEG, FEI, USA). The back scattered electron detector (BSED) and energy dispersive spectroscopy (EDS) were applied to inspect the microstructure and elemental distribution in the cross section of coating. The phase analysis was conducted by XRD (X'Pert Pro, PANalytical, Holland). The flexural strength of specimen was test by three points bending method with a span of 16mm and a test speed of 0.5mm/min. The apparent porosity was determined by the Archimedes displacement method, and the water absorption of coated specimens was tested by similar method after all the uncoated surfaces were sealed by epoxy (using polyamide resin as curing agent). The water absorption was defined by equation (1):

$$Water\,Absorption = \frac{W_2 - W_1}{W_0} \times 100\% \tag{1}$$

Where W_0 is the weight of dry specimen, W_1 is the weight of dry specimen with sealing material, and W_2 is the weight of wet specimen with sealing material.

The dielectric constant and loss were measured at 7-18GHz by the High Q Resonant Cavity method (Network Analyzer: E5071C, Agilent, USA; High Q Resonant Cavity Rig: University of Electronic Science and Technology, Chengdu, China) at room-temperature.

RESULTS AND DISCUSSION

Pretreatment of the Raw Powder

The Y_2O_3-Si2O-Al_2O_3 mixed powder was precalcined at 1300℃ in air for 4h. Figure 1 shows the XRD pattern of the Y-Si-Al-O powder. The β-$Y_2Si_2O_7$, Al_2SiO_5, and SiO_2 phases can be observed after the calcination which indicates that the oxides reacted with each other at 1300℃. After the calcination, the powder was ball milled for 4~5 days. The final Y-Si-Al-O powder particle size is in the range of 2~4μm and the morphology is shown in Figure 2.

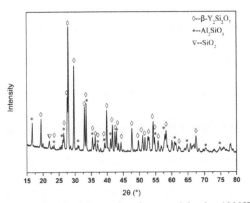

Figure 1. XRD pattern of Y-Si-Al-O powder after precalcined at 1300℃ for 4h in air.

Figure 2. Microstructure of Y-Si-Al-O powder after ball milled 4~5 days.

The calcination is necessary for two advantages: firstly, the precalcined powder performed better deposition features than the initial ternary oxide mixture since the solid content of Y-Si-Al-O slurry increased to 30 wt.% or above. As for the initial oxides slurry, the solid content can only reach the maximum of 20 wt.% where the slurry was too viscous to spray due to the bigger specific surface area resulted from smaller particle sizes, and the coatings tended to crack after drying; Secondly, the reactions leading to volume shrinkage may happen in advance before sintering of coatings on the substrate. All of these contribute to the densification of coating.

Phase Analysis of Y-Si-Al-O Coatings

The XRD patterns of Y-Si-Al-O coatings are shown in Figure 3. It shows that the C1350 is composed of β-$Y_2Si_2O_7$, mullite and SiO_2 phases. But the C1400 and C1450 display a totally different phase composition. As indicated in Figure 3, they are composed of β-Si_3N_4 and trace of glass phase. The result in C1350 can be interpreted as the Y-Si-Al-O powder reacted and crystallized on the surface of substrate, while, as for C1400 and C1450, the Y-Si-Al-O powder probably melted and infiltrated into the porous substrate completely during sintering as the increasing of sintering temperature, causing part of Si_3N_4 substrate was exposed and subsequently detected by the X-radiation.

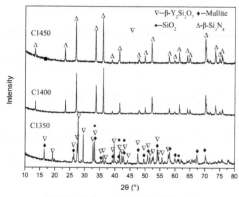

Figure 3. XRD patterns of Y-Si-Al-O coatings: (a) C1350, (b) C1400, and (c) C1450.

Macro- and Microstructure of Y-Si-Al-O Coatings

The heating and cooling rate were controlled at 2℃/min above 1300℃ to avoid cracking. After sintering, the flat and uniform grey coatings covered on the surface of all the substrate. Bubbles reported by S. Ramasamy et al.[11] for the Y_2O_3-SiO_2-Al_2O_3 coating system did not show up, because the protective N_2 atmosphere prevented the bubble-causing gaseous nitrogen oxide that formed during air sintering. Besides, the porous structure of substrate facilitated the bubble elimination by providing extra gas expel paths.

Figure 4 shows the SEM images of coating surface under different magnification. The coatings are uniform and crack-free, except for the C1350. As seen in Figure 4-(a), the surface of C1350 is porous with a lot of micron-scale holes, which is not suitable for moisture resistance purpose. While the relatively dense surface with small undulation can be observed for C1400 and

C1450 i.e. Figure 4-(b) and (c). As for C1350, the viscosity of liquid phase increased because of the lower sintering temperature, thus the mass transportation process was not as strong as C1400 and C1450. So the micron-scale holes formed in C1350.

More details of coating surface are showed in Figure 4-(d), (e) and (f). All the rod-like crystals were embraced by a layer of glass, resulting in the blur edges and terminals. The results suggested that the Y-Si-Al-O powders melted at the sintering temperatures and the viscous liquid enhanced the densification and consolidation process during sintering. Subsequently, the liquid phase filled up the spaces between grains after solidified by cooling down.

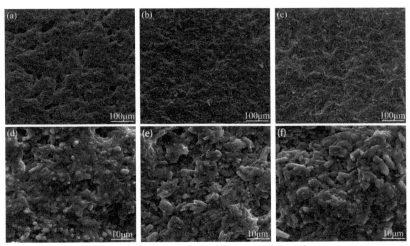

Figure 4. SEM images of Y-Si-Al-O coating surface:
(a), (d) C1350; (b), (e) C1400; (c), (f) C1450.

The cross sections of coatings were inspected by BSED. As seen in Figure 5, all the three coatings, with a thickness of about 100μm, are dense and bonded firmly on the porous substrate. The C1350 is different from C1400 and C1450 by means of having a definitely dual layer structure. The outer layer is brighter than the inner one, meaning that the outer layer contents more elements of larger atomic number (Z), in this case, which may be the yttrium or aluminum. As seen in Figure 5-(a), the bright crystal grains and dark grey intergranular phase formed a dense outer layer, and the inner layer is composed of light grey and dark grey phases. EDS section analysis was applied to identify elements composition on the four typical microstructures in C1350, as specified in Figure 6. Corresponding results are listed in Table 2. Section A contains element O, Si, and Y elements mainly, and the atomic ratio of the three elements is close to 3.5:1:1, which is the stoichiometric proportion of $Y_2Si_2O_7$. Thus, the bright grains in outer layer should be β-$Y_2Si_2O_7$. Section B is rich in O, Al and Si, which may be the mixture of mullite and SiO_2 phases. In the inner layer, section C mainly contents O and Si and small amount of Al and Y elements. It means that the light grey phase is possibly SiO_2 or Y-Si-Al-O glass. Section D is probably Si_3N_4 grain since it is only rich in N and Si elements. A conclusion can be drawn from all the results that the outer layer is composed of bright β-$Y_2Si_2O_7$ grains with mullite and SiO_2 as intergranular phases and the inner layer is the mixture of β-Si_3N_4 grains and Y-Si-Al-O glass formed by the infiltration of Y-Si-Al-O melts.

Figure 5. Cross section SEM (BSED) images of Y-Si-Al-O coatings:
(a) C1350, (b) C1400, (c) C1450.

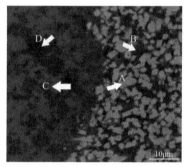

Figure 6. EDS section analysis on the four typical microstructures (A, B, C, D) in C1350.

Table 2. EDS section analysis results of the four typical microstructures in C1350.

Elements	A		B		C		D	
	wt.%	at.%	wt.%	at.%	wt.%	at.%	wt.%	at.%
N	\	\	0.33	0.52	2.32	4.00	30.47	45.63
O	33.36	63.41	46.55	64.44	36.03	54.41	6.72	8.81
Al	2.00	2.25	13.15	10.80	8.97	8.04	1.64	1.28
Si	16.51	17.87	26.46	20.87	32.68	28.11	58.39	43.62
Y	48.13	16.46	13.50	3.36	20.00	5.43	2.78	0.66

On the other hand, similar single layer microstructure for C1400 and C1450 are shown in Figure 5-(b) and (c). No bright β-$Y_2Si_2O_7$ crystal grains are observed in the coatings. The dense layers are composed of light grey and dark grey phases. EDS section analysis shows similar element composition with section C and D (Figure 6) in C1350 respectively. Thus the dense layers of C1400 and C1450 are mixture of Si_3N_4 substrate and Y-Si-Al-O glass as well. At 1400℃ and 1450℃, the melts have lower viscosity would infiltrate into the pores easier than the C1350.

With no massive residual liquid phase left on the surface, the single layer structure is formed for C1400 and C1450. Some liquid phase that was kept at the surface of the substrate by the force of capillary-action and immersional-wetting turned into glass and filled up the surface pores resulting in the dense surface of C1400 and C1450 after cooling down. The infiltration problem was also reported in the other coating systems for porous substrate.[8,9] It is unavoidable because the capillary action of the highly porous structure facilities the massive infiltration of liquid during sintering. The different cross section microstructure for the Y-Si-Al-O coatings is caused by the different degree of infiltration.

The Role of Y-Si-Al-O Coatings

The porosity and water absorption measured before and after coating are listed in Table 3. The porosity of substrate is slightly decreased after coating. However, the water absorption has been reduced significantly by 65.6-92.5%, which is attributed to the dense and uniform coating structure, especially for the C1400 (2.6%). The relatively high water absorption of C1350 (12.8%) may result from the micron-scale porous surface as shown in Figure 4-(a).

Table 3. The porosity and water absorption of substrate before and after coating.

Specimen	Substrates		Coatings	
No.	Porosity (%)	Water Absorption (%)	Porosity (%)	Water Absorption (%)
C1350	54.3	37.2	50.8	12.8
C1400	51.8	34.7	51.0	2.6
C1450	53.1	37.0	52.0	3.3

The flexural strength of substrate and substrate with coating is presented in Figure 7. The coating improves the flexural strength of the substrate for the C1350 and C1400. The flexural strength of C1400 is the highest (40.7±5.5MPa) which is improved by 15.6%, comparing to the substrate without coating (35.2±2.5MPa). The strength of C1450 (34.8±4.1MPa) is a little lower than the substrate. However, the variation is covered by the range of measurement error as shown in the figure. So the C1450 has only slight impact on the flexural strength of substrate.

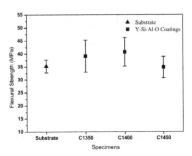

Figure 7. Flexural strength of substrate and substrate with Y-Si-Al-O coatings.

The dielectric constant and loss were measured at frequency range of 7-18GHz for each specimen before and after coating, and corresponding results are presented in Figure 8. The dielectric constant and loss are both increased after the coating. The dielectric properties of composite material depend on the intrinsic properties of material, structure, temperature and some other static dielectric constant. In this case, the porosity is an important structure parameter for the porous ceramics. The greater porosity results in a smaller dielectric constant. [6,7,12,13] So the degeneration of dielectric properties may be mainly attributed to decrease of the porosity after coating as seen in Table 3.

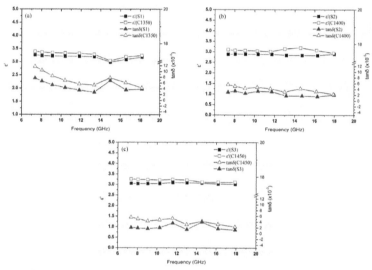

Figure 8 Dielectric constant and loss of substrate before and after coating. S1, S2, and S3 in the figures represent substrates for C1350, C1400 and C1450 respectively.

CONCLUSION

In summary, the Y-Si-Al-O environment barrier coatings were fabricated on the surface of porous Si_3N_4 substrate by slurry spray method. After sintered at 1350 ℃, 1400 ℃ and 1450℃, grey uniform and bubble-free coatings covered on the surface of substrates.

(1) Different coating phase composition and microstructure resulted from the different viscosity of the Y-Si-Al-O melts at each sintering temperature. Correspondingly, different cross section structure formed due to the different degree of liquid infiltration. Dense and uniform coatings were obtained after sintered at 1400℃ and 1450℃.

(2) The water absorption decreased as the density and uniformity of coating improved, and the water absorption of coating sintered at 1400℃ is as low as 2.6%.

(3) The flexural strength of substrate was improved by coatings except for the coating sintered at 1450℃. The degree of improvement is 15.6% for the coating sintered at 1400℃ with the value of 40.7±5.5MPa.

(4) The dielectric constant and loss of specimen increased due to the decrease of the

porosity after coating.

ACKNOWLEDGEMENT

This work was supported by the National Natural Science Foundation of China (Grant No. 90816018 and 51272206).

REFERENCES

[1] J. Barta, M. Manela, Si_3N_4 and Si_2N_2O for high performance radomes, *Mater. Sci. Eng.*, **71**, 256-272 (1985).

[2] Y. F. Xia, Y. P. Zeng, D. L. Jiang, Dielectric and mechanical properties of porous Si_3N_4 ceramics prepared via low temperature sintering, *Ceram. Int.*, **35**, 1699-1703 (2009).

[3] S. Q. Li, Y. C. Pei, C. Q. Yu, J. L. Li, Mechanical and dielectric properties of porous Si2N2O-Si3N4 in situ composites, *Ceram. Int.*, **35**, 1851-54 (2009).

[4] H. J. Wang, J. L. Yu, J. Zhang, D. H. Zhang, Preparation and properties of pressureless-sintered porous Si_3N_4, *J. Mater. Sci.*, **45**, 3671-3676 (2010).

[5] X. M. Li, L. T. Zhang, X. W. Yin, Effect of chemical vapor deposition of Si_3N_4, BN and B_4C coatings on the mechanical and dielectric properties of porous Si_3N_4 ceramic, *Scripta Mater.*, **66**, 33-36(2012).

[6] X. M. Li, X. W. Yin, L.T. Zhang, L. F. Chen, Y. C. Qi, Mechanical and dielectric properties of porous Si3N4-SiO2 composite ceramics, *Mater. Sci. Eng., A*, **500**, 63-69 (2009).

[7] R. B. Zhang, D. N. Fang, Y. M. Pei, L. C. Zhou, Microstructure, mechanical and dielectric properties of highly porous silicon nitride ceramics produced by a new water-based freeze casting, *Ceram. Int.*, **38**, 4373-4377 (2012).

[8] Q. Shen, Y. Yang, F. Chen, L. M. Zhang. Fabrication of Si3N4-based seal coating on porous Si3N4 ceramics, *Int. J. Mater. Prod. Technol.*, **42**(1-2), 12-20 (2011).

[9] C. C. Zhang, X. L. Li, H. M. Ji, Coating on porous Si3N4 based substrate with Sol-Gel slurry, *Integrated Ferroelectrics: An International Journal*, **138**, 111-116 (2012).

[10] N. Saito, K. Kai, S. Furusho, K. Nakashima, and K. Mori, Properties of nitrogen-containing Yttria-Alumina-Silica melts and glasses, *J. Am. Ceram. Soc.*, **86**[4] 711-716 (2003).

[11] S. Ramasamy, S. N. Tewari, K. N. Lee, R. T. Bhatt, and D. S. Fox. EBC development for hot-pressed Y2O3/Al2O3 doped silicon nitride ceramics, *Mater. Sci. Eng., A*, **527**, 5492-5498 (2010).

[12] J. Xu, D. M. Zhu, F. Luo, W. C. ZHOU, P. Li, Dielectric properties of porous Reaction-boned Si3N4 ceramics with controlled porosity and pore size, *J. Mater. Sci. Technol.*, **24**(2), 207-210 (2008).

[13] H. L. Du, Y. Li, C. B. Cao, Effect of temperature on dielectric properties of Si3N4/SiO2 composite and silica ceramic, *J. Alloys Compd.*, **503**, L9-L13 (2010).

CREEP AND ENVIRONMENTAL DURABILITY OF ENVIRONMENTAL BARRIER COATINGS AND CERAMIC MATRIX COMPOSITES UNDER IMPOSED THERMAL GRADIENT CONDITIONS

Matthew Appleby, Gregory N. Morscher
Department of Mechanical Engineering, The University of Akron, Akron, OH
Dongming Zhu
Durability and Protective Coatings Branch, NASA Glenn Research Center, Cleveland, OH

ABSTRACT

Interest in silicon carbide (SiC) fiber-reinforced silicon carbide ceramic matrix composites (CMCs) and environmental barrier coating (EBC) systems for use in high temperature structural applications has prompted the need for characterization of material strength and creep performance in complex aerospace turbine engine environments. Stress-rupture tests were performed on SiC/SiC composite systems with varying coating schemes to demonstrate material behavior under isothermal conditions. Additional testing was conducted under thermal stress gradients to determine the effect on creep resistance and material durability. In order to understand the associated damage mechanisms, emphasis was placed on experimental techniques as well as implementation of non-destructive evaluation (NDE), including electrical resistivity monitoring. The influence of environmental and loading conditions on life-limiting material properties is discussed.

INTRODUCTION

The high strength and temperature capability of ceramic matrix composites (CMCs) makes them a leading candidate for high temperature structural applications. Efforts to increase turbine engine efficiency have lead researchers to consider CMCs as a replacement for traditional superalloy components due to their potential for higher engine operating temperatures[1,2]. Continuous silicon carbide fiber-reinforced silicon carbide (SiC/SiC) CMCs have shown considerable promise in this field due to their mechanical properties and high temperature oxidation resistance, making them suitable for gas turbine engine hot-section components[3,4]. The use of CMCs for these applications combines the cost-saving benefits of reducing cooling requirements and weight, while increasing operating temperatures and engine output[5,6].

However, further investigation is required to determine the effect of thermo-mechanical loading conditions that more appropriately simulate real engine environments. Combustion section components are typically exposed to a high temperature environment with backside cooling, which results in a large thermal gradient across the thickness of the structure. Isothermal test conditions are therefore not an accurate representation of the given stress state and resulting damage and failure mechanisms. Accordingly, thermal gradient testing is preferred to measure creep rupture.

Silicon based ceramics, like those used in this work, exhibit high temperature oxidation resistance in dry environments due to the formation of a protective silica (SiO_2) layer. However, in combustion environments containing water vapor, the thermally grown SiO_2 layer can form volatile gas species (primarily $Si(OH)_4(g)$), leading to significant and rapid recession of the component[7]. For this reason, environmental barrier coatings (EBCs) have been developed to protect silicon based materials from high temperature corrosion. Coating stability is dictated primarily by material selection, operating conditions, and thermo-mechanical loading[8-10]. In this study, the effect of advanced EBCs on SiC/SiC CMCs under laboratory imposed thermal gradient and isothermal creep conditions was considered. Particular interest was paid to their effect on stress rupture life, stressed-oxidation resistance and CMC degradation mechanisms.

Before SiC/SiC composites can be successfully implemented in high temperature structural applications, it is necessary to characterize and monitor damage associated with high temperature loading. Due to the processing method, melt-infiltrated (MI) SiC/SiC composites contain an excess of free silicon, which provides good electrical conductivity. In the past, many researchers have demonstrated the use of electrical resistivity (ER) measurements as a damage sensing technique for carbon-fiber reinforced polymers (CFRP) composite systems[11-19]. More recent studies have shown success in correlating SiC/SiC composite damage with increased stress under room temperature tensile loading[20], and under isothermal tensile creep rupture testing[21]. The work discussed here extends the use of electrical resistance measurements to include thermal gradient tensile creep conditions in order to monitor material degradation and understand material behavior under more realistic testing environments.

EXPERIMENTAL MATERIALS AND METHODS

Materials
     The composite specimens used for this study were comprised of eight plies of symmetric 0/90 layups of 2D woven five harness satin fiber preforms. The preforms were coated with a thin chemical vapor infiltration (CVI) boron nitride (BN) interphase layer, followed by a layer of CVI SiC, slurry cast SiC, and final densification by melt-infiltration (MI) of molten silicon. The composite samples were manufactured by Goodrich Aerospace and the reinforcing fibers used in these composites were Tyranno ZMI fibers produced by Ube Industries of Japan. Test specimens were machined from 4 mm thick panels to produce contour dog-bone shapes with an overall length of 152 mm, a gauge length of 25 mm, and grip and gauge widths of 12.6 mm and 10 mm respectively. The length of the specimen was aligned along the 0° fiber direction. As previously mentioned, this study includes the evaluation of two NASA turbine engine environmental barrier coating (EBC) systems, which are listed below in Tables I and II. In general, these coatings consisted of a thin HfO_2-Si bond coat layer for adhesion, followed by a ceramic topcoat layer for environmental protection[22]. Both the bond coat and the ceramic topcoat were deposited via electron beam physical vapor deposition (EB-PVD). The bond coat was deposited as a mixture of two materials (Si and HfO_2), and the topcoat was a single material. Hafnium-based coatings were chosen for their high melting temperatures, low thermal conductivity and good thermal and environmental stability. For the two coated samples, the coatings were deposited on one face and both sides of the dog-bone specimens in order prevent heating along the edge of the substrate.

Table I. Test sample specifications.

	Gauge Width (mm)	Thickness (mm)	Fiber Volume Fraction	Coating
ZMI-0	10.042	4.191	0.272	Uncoated
ZMI-1	10.129	4.231	0.274	Uncoated
ZMI-2	10.455	4.686	0.274	EBC-1
ZMI-3	10.686	4.342	0.298	EBC-2

Table II. Environmental Barrier Coating specifications.

	Bond Coat	**Ceramic Coating** (layer thickness)
EBC-1	HfO_2-Si	$Yb_2Si_2O_7$ (~ 127 μm) + Hf-RE silicate (~254 μm)
EBC-2	HfO_2-Si	HfYb silicate (~ 381 μm)

High Temperature Testing

The elevated temperature isothermal tensile creep tests were performed using a screw-driven test machine (Instron Model 5569) and a $MoSi_2$ element, resistance-heated furnace located between the top and bottom grips. Prior to tensile loading for creep testing, samples were held at the target temperature for 20 minutes to ensure consistent heating. Strain was measured by a high temperature extensometer with knife-edged SiC rods in contact with the 25 mm gauge section of the sample. If the sample was capable of withstanding the creep conditions for over 500 hours it was unloaded and immediately reloaded at temperature with a constant cross-head loading rate of 0.127mm/min until fracture in order to measure retained material properties.

Thermal gradient tensile creep tests were performed using custom built rig at the NASA Glenn Research Center (Cleveland, OH) capable of applying through-thickness thermal gradients across the EBC/CMC systems (see Figure 1 for details). While the specimen is free to extend in the loading direction, any bending due to expansion caused by the temperature gradient is restrained by the fixed ends. Assuming a linear temperature gradient, this will induce a compressive thermal stress on the heated surface and a corresponding tensile stress on the backside. This is in contrast to the isothermally tested sample in which uniform heating leaves it subjected to the tensile creep load only.

The specimen surface was heated by a 3.5 kW high-heat flux CO_2 laser system, which is described in better detail in previous works[23,24]. Using an integrated rotating lens, a uniform laser power distribution is achieved over a 38 mm region of the sample face. In order to induce the desired through-thickness thermal gradient, controlled cooling air was delivered to the backside of the CMC. The front and backside temperatures of the specimen were monitored during testing by an 8μm and two-color pyrometers, respectively. Once the material surface reached the desired test temperature the creep load was applied via the hydraulic load cell. Similar to the isothermal tensile creep tests, the samples were fast fractured (at temperature) if creep rupture had not occurred after 500 hours in order to determine the remaining material strength.

 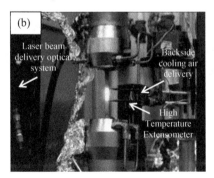

Figure 1. (a) Laser High heat-flux thermal gradient creep test rig at NASA GRC, (b) close-up of dog-bone specimen under tensile creep testing.

Electrical Resistance Measurements

In-situ electrical resistance monitoring was incorporated into the mechanical testing rigs described above in order to observe and characterize the accumulated damage. To increase ER measurement sensitivity, a four-point probe measurement technique was used. The ends of the specimens were wrapped in copper mesh to act as ER measurement electrodes as well as prevent uneven grip pressure during sample loading. A micro-ohm multimeter (Model 34420A, Agilent) was used to supply a constant DC electrical current and measure the potential difference along the sample. Alumina (Al_2O_3) wedge grip-inserts were used to grip the specimens and prevent electrical shorting between the ER electrodes. This work is the first time ER measurement has been used in high heat-flux laser tensile creep testing of CMCs.

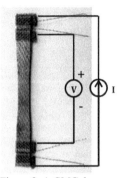

Figure 2. A CMC dog-bone specimen with attached ER electrodes.

In order to determine the validity and sensitivity of the ER monitoring technique, it was necessary to understand the contribution that the heated section of the specimen had on the electrical measurement of the entire tensile sample. Figure 3 depicts the unique NASA GRC laser heating setup which was used to characterize the electrical resistivity of the CMC specimens as a function of gauge temperature. The dog-bone specimen was not mechanically loaded during the test, which allowed for evaluation electrical resistance solely as a function of temperature. The specimen was mounted between a stainless-steel plate and an aluminum aperture plate to prevent heating along the edge and sides of the sample. To reduce specimen heat loss through the fixture, Al_2O_3 rods were placed above and below the specimen to provide air gaps between the mounting plates. The front and backside temperatures of the heated section were monitored using infrared pyrometers, and an R-type thermocouple was mounted at the end of the sample.

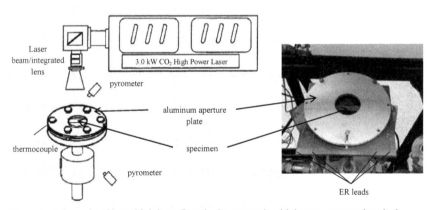

Figure 3. Schematic of laser high heat-flux rig for measuring high temperature electrical resistivity of CMCs. The typical uniform heated section length was 30-40 mm.

RESULTS AND DISCUSSION

Electrical Response to Heating

To investigate the temperature dependence of ER, an uncoated SiC/SiC sample (ZMI-0) was monitored during heating and cooling. The results of the electrical resistivity change during heating cycles are shown below in Figure 4. The laser power was increased gradually until the surface temperature in the gauge section reached a maximum temperature of approximately 1000°C. A clear quantitative increase in electrical resistivity is measured as the temperature of the CMC is increased. A second cycle demonstrated good agreement with the initial heating, although some scatter in the data was observed. However, the obvious trend in the data was consistent which may make this technique useful for the evaluation and modeling of high temperature CMC electrical properties.

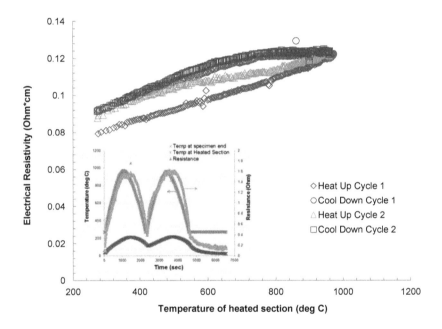

Figure 4. Plot of electrical resistivity of ZMI-0 sample with increasing temperature. The inset shows heating cycles used in testing.

Tensile Creep Behavior

Table III summarizes the creep properties, elastic moduli, electrical properties and retained strength properties of each experiment. The first laser-heated sample (ZMI-1) was uncoated and exposed to 1100°C at the CMC surface, with a recorded CMC backside temperature of 1034°C. The coated laser-heated sample (ZMI-2) was heated to a temperature of 1282°C at the EBC surface, and a temperature of 1010°C was maintained on the backside of the CMC substrate. It was assumed that if the CMC backside temperatures were equivalent, then the CMC surface temperatures should also be the same. Therefore the cooling air for the laser-based creep tests was adjusted to maintain similar backside temperatures and hence similar through-thickness temperature gradients. The isothermal temperature condition of 1050°C was selected to be comparable to the average CMC temperature of the thermal gradient tests.

Table III. Results of creep testing and associated properties. The test temperatures refer to the furnace temperature for the isothermal test, and the front/backside temperatures of the heated region of the laser-tested samples.

Test Sample	Coating	RT Resistivity $(\sigma = 0)$	Resistivity at T $(\sigma = 0)$	Temperature	Stress	E_0 at T	Time	Total Strain	Creep Strain	Post Creep Properties			
										Resistivity at T $(\sigma = 0)$	E at T	σ_{UTS} at T	Strain
		(Ohm*cm)	(Ohm*cm)	(°C)	(MPa)	(GPa)	(hrs)	(%)	(%)	(Ohm*cm)	(GPa)	(MPa)	(%)
ZMI-1*	Uncoated	0.0367	0.0935	1100/1034	68.9	209	127	0.13237	0.04782	0.189	-	-	-
ZMI-2*	EBC-1	0.0636	0.113	1282/1010	68.9	-	526	-	-	0.167	-	115	-
ZMI-3	EBC-2	0.0648	0.130	1050	68.9	153	501	0.07902	0.03167	0.183	173.6	138	0.1414

*Laser Test

A number of interesting observations arise from the tensile creep data. First, ZMI-1 is the only specimen not capable of surviving the full 500 hour creep test length (failing after only 127 hours). Secondly, while ZMI-2 and ZMI-1 were exposed to similar thermo-mechanical loading across the CMC substrate, ZMI-2 was capable of lasting 526 hours under these conditions and displayed a significantly lower change in ER. The tensile creep curves are plotted in Figure 5 as the total strain (i.e. creep strain plus loading strain) versus time, along with the associated normalized electrical resistance change (change in resistance normalized by the initial resistance at temperature) versus time. The ZMI-2 sample data is not shown due to a malfunction with the high temperature extensometer.

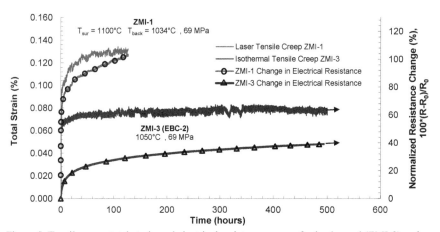

Figure 5. Tensile creep total strain and electrical resistance curves for isothermal (ZMI-3) and thermal gradient (ZMI-1) conditions. Note that the data for the specimen that did not rupture is indicated by arrows.

Note the considerable difference in the strain and electrical behavior between ZMI-1 and the isothermally heated specimen (ZMI-3). The ZMI-1 sample exhibits a larger degree of primary creep and electrical resistance change. While the creep curve shows a decreasing strain rate over time, a clear steady state creep regime (like the one seen in the ZMI-3 curve) was never achieved. The increase in electrical resistance during these creep tests is likely caused by the accumulation of matrix microcracking that limits the flow of electrical current through the composite.

Several micrographs were taken of the fracture surface in order to investigate the possible differences in failure mechanisms of the three specimens. Figure 6 shows the fracture surface of the uncoated CMC sample (ZMI-1) exposed to a thermal gradient with laser heating. The fracture surface appears quite smooth and exhibits little to no fiber pull-out. There was evidence of oxidation across the entire fracture surface, including the matrix surface, the BN interphase, and fiber fracture surfaces. The higher magnification images in Figure 6 (labeled as 1 and 2) show that the oxidation of the fiber fracture surfaces was prevalent throughout the fracture surface of the bulk specimen. The fact that the fiber fracture surfaces were oxidized indicates that these fibers failed some time before the final fracture of the composite. This failure is most likely due to intrinsic fiber degradation or local stress concentrations that arise from local load-sharing of fibers which were strongly bonded together by oxidation products. Oxidation processes are assumed to be the main contributor to the premature stress-rupture of this sample. Therefore, the life of this CMC could have been prolonged by an environmentally protective coating.

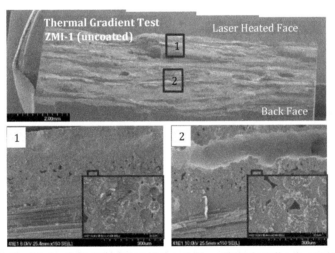

Figure 6. SEM micrographs of fracture surface of uncoated laser-heated tensile creep test ZMI-1.

Figure 7 shows a portion of the ZMI-1 gauge section that was cut and polished along its edge in order to observe the extent of matrix cracking in the sample. ZMI-1 showed a number of unbridged transverse matrix cracks (highlighted in the figure by dashed lines) that penetrated from either surface several plies deep, with some cracks that extended completely through the thickness of the composite. These matrix cracks allowed for internal exposure of the composite to the oxidizing environment. Oxygen that penetrated into these cracks reacted with the BN and SiC to form borosilicate glass that can strongly bond fibers to each other and to the matrix. These strongly

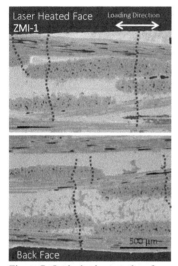

Figure 7. Optical micrographs of polished longitudinal section of ZMI-1. The dashed lines highlight unbridged matrix cracks.

bonded fibers generate local load-sharing conditions that created stress concentrations that propagated the crack further into the composite. These observations suggest that failure occurred by oxidation-assisted crack growth.

The fracture surface of the coated thermal gradient sample ZMI-2 is shown in Figure 8. In contrast to ZMI-1, most of the observed oxidation in ZMI-2 is along the back surface of the composite. The fracture surface also exhibited a high degree of fiber pull-out near the coated surface and significantly less near the backside. This suggests that the EBC acted to reduce the amount of oxygen ingress into the matrix from the coating/substrate interface, resulting in significantly less oxidation-assisted crack growth. The addition of the EBC reduced oxidation ingress on the heated surface, preventing strong bonding of the reinforcing fibers. The weak interfacial bonding allows for debonding between the fiber/BN interface or the BN/matrix interface. This debonding enabled global load-sharing amongst the fibers in the presence of a brittle matrix crack, leading to increased fracture toughness and longer creep life. This behavior was indicated by the fibers being "pulled out" of the matrix fracture surface. To show the representative matrix cracking associated with this sample, a section of ZMI-2 is shown in Figure 9. While both laser-heated CMCs saw similar thermally induced stress gradients, the majority of the cracking began on the uncoated surface, while only a small number of cracks propagated to (or originated from) the front surface of the sample. The tensile thermal stress generated by the temperature gradient, in addition to the tensile creep load applied to the sample, could be the reason for the preferential crack initiation on the backside. The

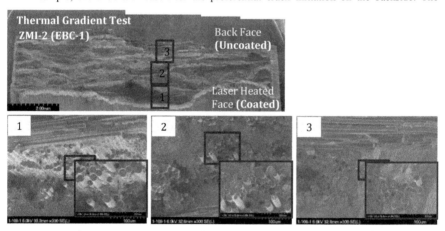

Figure 8. SEM micrographs of the fracture surface of EBC-1coated laser-heated sample ZMI-2. Higher magnification images show increased fiber pull-out near EBC/CMC interface.

transverse matrix cracks seen in ZMI-2 were not through thickness and in most cases did not penetrate past the second or third ply of the composite. While Figure 9 shows some level of EBC cracking, bond coat delamination leading to coating spallation at the EBC/CMC interface was not observed. The heated surface retained an intact EBC with few matrix cracks and a higher degree of fiber pull-out than ZMI-1, which reinforces that the EBC reduces oxygen ingress into the matrix. The protection from the environment decreases the magnitude of oxidation-assisted cracking in the composite, leading to an increase in creep resistance. The micrographs showing the fracture surface associated with the coated isothermal tensile creep sample (ZMI-3), are shown in Figure 10. The higher magnification micrographs show representative regions across the thickness of the specimen. ZMI-3 shows considerably more fiber pull-out than either of the laser tested (i.e. gradient heated) samples. The only oxidation that was observed was in the first 0° ply of the backside of the CMC. As previously mentioned, higher levels of fiber debonding and sliding lead to the pseudo-plastic behavior desired for brittle composites. This is evident by the high post-creep strength and strain capability of this composite.

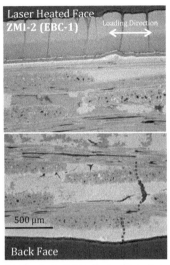

Figure 9. Optical micrographs of longitudinal section of ZMI-2. This sample showed mainly backside cracking that did not penetrate through thickness.

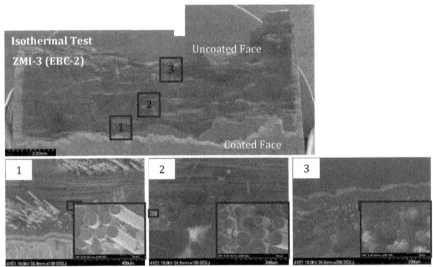

Figure 10. SEM micrographs of the fracture surface of EBC-2 coated isothermal sample ZMI-3. The high degree of fiber pull-out and low level of fracture surface oxidation are shown in the higher magnification images 1-3.

Figure 11 shows the matrix cracking associated with ZMI-3. While there was some unbridged matrix cracking, it was quite limited and did not penetrate past the first ply of the composite. As previously suggested, the small number of matrix cracks limited oxidation effects and lead to an increase in creep resistance. The micrographs show that the isothermally heated sample suffered less severe matrix microcracking and environmental attack than the gradient heated specimens. It is clear that damage accumulation under isothermal test conditions is not the same as damage seen in tests performed with a thermal gradient. This indicates the need for testing that more accurately simulates engine conditions to gain a proper understanding of the failure that will be encountered.

Figure 11. Optical micrographs of longitudinal section of ZMI-3. This isothermally tested sample showed only a small number of transverse matrix cracks.

CONCLUSION

The coupled ER measurements were shown to be sensitive to both temperature and damage state of the material, leading to confidence in its use as a non-destructive evaluation technique. There was a significant difference in creep life and associated failure mechanisms of CMC materials due to the presence/absence of EBC systems and thermal gradient conditions. Higher levels of matrix cracking were observed in CMC substrates exposed to thermal stress gradient conditions the sample exposed to isothermal conditions. As these cracks open, oxidation ingress increases, leading to premature failure of the material due to increased bonding between the fiber and matrix. The addition of the EBC demonstrated a significant reduction in oxidation, which allowed for fiber/matrix debonding and subsequently longer creep life. The exposed backside surfaces of the EBC coated specimens appeared to be the main source of oxidation ingress due to an increased tensile stress state (under thermal gradient conditions) and exposure to the environment. The isothermally tested sample exhibited the highest degree of fiber pull-out and lowest degree of matrix cracking, leading to the greatest creep tolerance. This phenomenon demonstrates one of the possible benefits of EBCs, and the detrimental effect of thermal gradients on creep properties. Of course continued testing is required to assure repeatability, and further investigation into the mechanisms and environmental contributions associated with premature failure is crucial to the understanding of precise material failure modes.

ACKNOWLEDGEMENT

Many thanks to the NASA Glenn Research Center for their financial support under the Graduate Student Researchers Program (Grant No. NNX11AL03H), and the support of my mentors and colleagues Dr. Zhu and Dr. Morscher.

REFERENCES

[1]D. Brewer, "HSR/EPM Combustor Materials Development Program," *Mater. Sci. Eng.*, A261 [1–2] 284–291 (1999).
[2]G. S. Corman, et al., "Rig and Engine Testing of Melt Infiltrated Ceramic Composites for Combustor and Shroud Applications," *J. Eng. Gas Turbines Power*, 124 [3] 459–464 (2002).

[3]G.N. Morscher and V. Pujar, "Creep and Stress-Strain Behavior After Creep for SiC Fiber Reinforced, Melt-Infiltrated SiC Matrix Composites" *J. Am. Ceram. Soc.* 89 [5] 1652-1658 (2006).

[4]Dicarlo, J. A., Yun, H. M., Morscher, G. N. and Thomas-Ogbuji, L. U. Progress in SiC/SiC Composites for Engine Applications, in High Temperature Ceramic Matrix Composites (eds W. Krenkel, R. Naslain and H. Schneider), Wiley-VCH Verlag GmbH & Co. KGaA, Weinheim, FRG (2006).

[5]J. A. DiCarlo, H. -M. Yun, G. N. Morscher, and R. T. Bhatt, "SiC/SiC Composites for 1200°C and Above," *Handbook of Ceramic Composites.* ed., N. P. Bansal. Spinger, Berlin, 78–98, 2005.

[6]M. C. Halbig, M. H. Jaskowiak, J. D. Kiser, and D. Zhu, "Evaluation of Ceramic Matrix Composite Technology for Aircraft Turbine Engine Applications", 51st AIAA Aerospace Sciences Meeting including the New Horizons Forum and Aerospace Exposition, AIAA 2013-0539, 2013.

[7]E. J. Opila, J. L. Smialek, R. C. Robinson, D. S. Fox, and N. S. Jacobson, "SiC Recession Caused by SiO_2 Scale Volatility under Combustion Conditions: II, Thermodynamics and Gaseous-Diffusion Model," *J. Am. Ceram. Soc.*, 82 [7] 1826-34 (1999).

[8]Kang N. Lee, Dennis S. Fox, Jeffrey I. Eldridge, Dongming Zhu, Raymond C. Robinson, Narottam P. Bansal, and Robert A. Miller, "Upper Temperature Limit of Environmental Barrier Coatings Based on Mullite and BSAS", Journal of the American Ceramic Society, 86 (2003), pp. 1299-1306.

[9]D. Zhu, K. N. Lee, and R. A. Miller, "Thermal Conductivity and Thermal Gradient Cyclic Behavior of Refractory Silicate Coatings on SiC/SiC Ceramic Matrix Composites," *Ceram. Eng. Sci. Proc.*, vol. 22, pp. 443-452, 2001.

[10]D. Zhu and R.A. Miller, "Thermal and Environmental Barrier Coating Development for Advanced Propulsion Engine Systems", 48th Structures, Structural Dynamics, and Materials Conference sponsored by the AIAA, ASME, ASCE, AHS, AIAA Paper 2007-2130.

[11]J. C. Abry, S. Bochard, A. Chateauminois, M. Salvia, and G. Giraud, "In Situ Detection of Damage in CFRP Laminates by Electrical Resistance Measurements," *Compos.* Sci. Tech., 59 925–935 (1999).

[12] D. -C. Seo and J. -J. Lee, "Damage Detection of CFRP Laminates Using Electrical Resistance Measurement and Neural Network," *Compos. Struct.,* 47 [1–4] 525–530 (1999).

[13]S. Wang, D. D. L. Chung, and J. H. Chung, "Impact Damage of Carbon Fiber Polymer–Matrix Composites, Studied by Electrical Resistance Measurement," *Composites*, A36 1707–1715 (2005).

[14]M. Kupke, K. Schulte, and R. Schuler, "Non-Destructive Testing of FRP by d.c. and a.c. Electrical Methods," *Compos. Sci. Technol.*, 61 837–847 (2001).

[15]I. Weber and P. Schwartz, "Monitoring Bending Fatigue in Carbon-Fibre/Epoxy Composite Strands: A Comparison Between Mechanical and Resistance Techniques," *Compos. Sci. Technol.*, 61 [6] 849–853 (2001).

[16]J. C. Abry, Y. K. Choi, A. Chateauminois, B. Dalloz, G. Giraud, and M. Salvia, "In-Situ Monitoring of Damage in CFRP Laminates by Means of AC and DC Measurements," *Compos. Sci. Technol.*, 61 855–864 (2001).

[17]Y. Okuhara and H. Matsubara, "Memorizing Maximum Strain in Carbon-Fiber-Reinforced Plastic Composites by Measuring Electrical Resistance Under Pre-Tensile Stress," *Compos. Sci. Technol.*, 65 [14] 2148–2155 (2005).

[18]A. Todoroki, K. Omagari, Y. Shimamura, and H. Kobayashi, "Matrix Crack Detection of CFRP Using Electrical Resistance Change with Integrated Surface Probes," *Compos. Sci. Technol.*, 66 [11–12] 1539–1545 (2006).

[19]S. Wang and D. D. L. Chung, ''Self-Sensing of Flexural Strain and Damage in Carbon Fiber Polymer-Matrix Composite by Electrical Resistance Measurement,'' *Carbon*, 44 [13] 2739–2751 (2006).

[20]C. E. Smith, G. N. Morscher, and Z. H. Xia, ''Monitoring Damage Accumulation in Ceramic Matrix Composites Using Electrical Resistivity,'' *Scripta Materialia*, 59 [4] 463–466 (2008).

[21]Smith, C. E., Morscher, G. N. and Xia, Z., "Electrical Resistance as a Nondestructive Evaluation Technique for SiC/SiC Ceramic Matrix Composites Under Creep-Rupture Loading," *International Journal of Applied Ceramic Technology*, 8: 298–307 (2011).

[22]D. Zhu, and Milelr, "Multi-functionally graded environmental barrier coatings for Si-based ceramic components," U.S. Patent No. 7,740,960 B1, 2010.

[23]Dongming Zhu and Robert A. Miller, "Determination of Creep Behavior of Thermal Barrier Coatings under Laser Imposed High Thermal and Stress Gradient Conditions," *Journal of Materials Research*, 14 (1999), pp. 146-161.

[24]D. Zhu, Dennis S. Fox, Louis J. Ghosn and Bryan Harder, "Creep Behavior of Hafnia and Ytterbium Silicate Environmental Barrier Coating Systems on SiC/SiC Ceramic Matrix Composites", The International Conference on Advanced Ceramics & Composites, January 23-28, 2011.

DYNAMIC OBLIQUE ANGLE DEPOSITION OF NANOSTRUCTURES FOR ENERGY APPLICATIONS

G.-C. Wang[+], I. Bhat[++], and T.-M. Lu[+]
[+]Department of Physics, Applied Physics and Astronomy
[++]Department of Electrical, Computer and Systems Engineering
Rensselaer Polytechnic Institute
110 8th St., Troy, NY 12180

ABSTRACT

The century old oblique angle deposition (OAD) of porous films has received renewed attention recently. The technique can tailor not only the morphology, but also the texture of nanostructures. Nanostructures with a variety of morphologies and textures can be achieved by engineering the incident flux angle, substrate rotation speed, and rotation axis. These different deposition configurations can change the shadowing effect and the texture selection rule dynamically during the nanostructure self-assembly process. These artificially fabricated nanostructures have been used as sensors in many applications. Among the textured films, a particular class of biaxial films such as MgO has been used as a buffer layer to grow, for examples, high temperature superconductors and ferroelectrics. More recently, it has been shown that biaxial CaF_2 nanorods can be coated on amorphous substrates at room temperature using OAD. We demonstrated that single crystal-like CdTe can be grown by metal organic chemical vapor deposition (MOCVD) directly on CaF_2 on glass substrate despite a large lattice mismatch between CdTe and CaF_2. We also showed that single crystal-like Ge can be epitaxially grown on biaxial CaF_2 buffer on glass. These films have potential applications as low cost photovoltaic devices. We will also introduce a new oblique angle flipping rotation mode to grow biaxial metallic nanorods such as W and Mo. These films may also find applications as buffers to grow functional materials for energy applications.

INTRODUCTION

High efficiency photovoltaic devices normally have been fabricated from single crystalline substrates [1]. In practice, these single crystalline substrates are expensive and large scale production for widespread usage has not been realistic. Instead, large volume production of solar cells is on less expensive non-crystalline substrates such as glass. Typically the films grown on glass are polycrystalline [2]. The energy conversion efficiency and the long term stability of these polycrystalline film solar cells are not as good as that of the single crystalline counterpart due to the presence of the randomness of the grain boundaries.

A challenging scientific question can be raised. Can one grow single crystal films on top of a glass substrate by physical or chemical vapor deposition? The answer is unlikely because there is no regular lattice on a glass surface that can induce an epitaxial growth of a single crystal film. A schematic in Fig. 1(a) shows a polycrystalline film with small size grains and grain boundaries on amorphous substrate. Recently, an alternative approach has been suggested that a dramatic gain in the conversion efficiency may be achieved if one uses a biaxially oriented buffer layer on glass (see Fig. 1(b)) to grow biaxial semiconductor films for solar cell devices compared to that of films grown directly on glass. Biaxial films are not exactly single crystal but have strongly preferred crystallographic orientations in both the out-of-plane and in-plane directions [3-5]. For example, the in-plane angular misorientation between grains can be a few degrees (small angle). In this sense, biaxial films can be considered as "quasi single crystals". See Fig. 1(c). Biaxial buffer layers have been successfully used to grow high T_c superconductor films to achieve a high

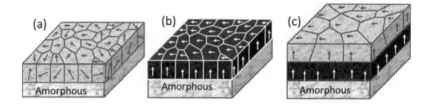

Figure 1. (a) A polycrystalline film with small grains and grain boundaries on an amorphous substrate, (b) a biaxial film on an amorphous substrate, and (c) a near single crystal on a biaxial buffer layer on an amorphous substrate.

critical current [6]. It was suggested that high quality semiconductor films for solar cell devices with biaxial crystal orientation may be fabricated if biaxial buffer layers are used [7, 8]. There is evidence that the carrier recombination rate at small angle grain boundaries can be low compared to that of random grain boundaries [9, 10]. It has been shown that for small angle grain boundaries below 4°, the trap density of Si is dramatically reduced and the film can lead to a much higher carrier mobility [10]. This is very encouraging in that the biaxial films on glass may provide high solar conversion efficiency with low cost.

One of the most promising strategies to grow biaxially oriented buffer layers on glass is by oblique angle deposition (OAD), also called incline substrate deposition [11-13]. OAD deposition is a simple, inexpensive, and scalable technique for creating biaxially oriented films. An example of the biaxial films grown by oblique angle deposition is MgO, which had been used as the buffer layer for the growth of oriented high T_c superconductor films [12, 13]. Besides OAD there are other methods to grow biaxial films. These include rolling-assisted template technique (named RABiTS--Rolling Assisted Biaxially Textured Substrates, or cube-texture substrates) [14-19] and ion-beam-assisted deposition (IBAD) [7, 20, 21].

In the present work, we will summarize our recent work on the creation of biaxial CaF_2 buffer layer using the OAD technique on glass. CaF_2 has been known as an excellent buffer layer material for the growth of a variety of semiconductor materials including Si [22-27] and compound semiconductors such as CdTe or GaAs [26-32]. In the past, high quality single crystal CaF_2 buffer layers have been achieved on single crystal Si substrates. Here we adopted the OAD process to grow biaxially oriented CaF_2 films on amorphous substrates as buffer layers. These buffer layers are used to grow biaxially oriented semiconductor films such as Ge and CdTe by physical vapor deposition and metalorganic chemical vapor deposition (MOCVD), respectively at a temperature below the glass melting temperature of ~600 °C. Our experiments showed that the optimum growth temperature of Ge and CdTe on CaF_2 buffer layer is <450 °C. For other semiconductors such as Si, one would require a temperature higher than or close to the glass melting temperature for biaxial growth. We have also developed a new flipping rotation mode in dynamic OAD to grow biaxial metal films of Mo and W on amorphous substrates. These films can serve as conducting buffer layers to grow epitaxial films.

We characterized the morphology of films using a field emission scanning electron microscopy (SEM, Carl Zeiss Supra 1550), structure and texture using X-ray pole figure analyses with an area detector (Bruker D8 Discover diffractometer, wavelength = 0.15405 nm), and near surface structure and texture determination using reflection high energy electron diffraction (RHEED) surface pole technique developed in our lab. The principle, measurement, and construction of such RHEED surface pole figures have been described in details [4, 33]. Transmission electron microscopy (TEM) was used to study the grain boundary in Ge and CdTe films and the interfaces of Ge/CaF$_2$ and CdTe/CaF$_2$ films.

EXPERIMENTAL

Texture classification
    Polycrystalline or nanocrystalline grains in a thin film often have a certain preferred direction that is developed during growth. This preferred orientation of the crystals is called the texture [3]. There are two extreme cases. In the random orientation case, the grains in the film have no preferred orientation, as shown in Fig. 2(a). If all the grains are oriented in the same direction, both the out-of-plane and in-plane directions with respect to the substrate, the film is basically a single crystal as shown in Fig. 2(d). To grow a single crystal film, one typically requires a single crystal substrate (and perhaps at an elevated temperature). In between these two extreme cases, there are many other possibilities.

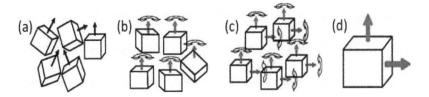

Figure 2. Various crystal orientations: (a) random orientation, (b) one-degree orientation or fiber, (c) two-degree orientation or biaxial, and (d) single-crystal orientation. The curved double arrows indicate deviations from the preferred directions (blue and red arrows) or dispersion. From [3].

    The most common one is a fiber texture structure, or the one-degree orientation (I-O), as shown in Fig. 2(b). One-degree orientation refers to a texture in which one crystallographic axis of the crystals points in a particular preferred direction labeled as <hkl>-I-O. Often, this direction is oriented normal to the substrate (out-of-plane) if the deposition flux is incident normally to the substrate. This I-O orientation is often not perfect in the sense that the <hkl> orientation needs not be exactly normal to the surface but can have a dispersion around the surface normal. The grains have no preferred orientation in the plane of the substrate (in-plane) and are completely random. If the deposition flux is incident upon the substrate with an oblique angle α with respect to the surface normal, one can have another type of texture called biaxial texture or two-degree orientation (II-O). See Fig. 2(c). In this case, the two crystallographic axes point in two preferred directions labeled as <h$_1$k$_1$l$_1$><h$_2$k$_2$l$_2$>-II-O. The <h$_1$k$_1$l$_1$> direction is basically a tilted I-O direction. The second direction, <h$_2$k$_2$l$_2$>, is an in-plane preferred orientation. Since the incident flux angle α can be any value, there are infinite possible arrangements for the II-O orientation. Furthermore, as we will see later, in OAD one can dynamically vary the incident flux angle and/or substrate rotation speed to generate novel and rich texture structures that cannot be obtained by any other means.

X-ray pole figure analysis of textures

The most powerful technique for studying the texture of a polycrystalline or nanocrystalline film over a large area is by diffraction. In particular, the x-ray pole figure technique has been the most popular technique for film-texture analysis. The x-ray can probe the bulk of the thin film, and the texture information obtained is an average texture of the entire film. A pole figure represents the intensity profile from a specific Bragg angle (one family of planes), as the sample is rotated azimuthally in the ϕ direction around the surface normal and tilted out of plane in the χ direction. Conventionally, a point detector is used and is parked at a specific Bragg angle to collect data for analyzing a particular hkl plane family. Usually, a θ-2θ scan is performed to identify the hkl planes before the pole figure measurements are taken [34].

Figures 3(a), 3(b), 3(c) and 3(d) show schematics of pole figures representing the textures described in Figs. 2(a), 2(b), 2(c) and 2(d), respectively. The pole figure of a biaxial texture would be in between Figs. 2(c) and 2(d). The regions with a higher density of small circles indicate higher diffraction intensity. The χ scan passes through the origin of the pole figure and the ϕ scan is the azimuthal scan in the sample plane. For more sophisticated textures, the pole figures are more complex and with detailed analysis one can obtain an average texture of the films.

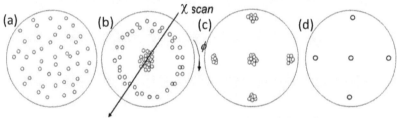

Figure 3. (a), (b), (c) and (d) are schematic diagrams showing examples of the pole figures representing the texture orientation described in Figs. 2(a), 2(b), 2(c) and 2(d), respectively. The regions with high density of small circles indicate higher diffraction intensity.

Oblique-angle deposition and physical shadowing effect

The growth of biaxial texture is a result of the shadowing effect in oblique angle deposition (OAD), a subject of keen interest from both fundamental and practical point of views. During shadowing growth, the direction of incident flux can have a profound impact not only on morphology but also on the evolution of the crystal orientation of the film. This is exciting in that very few techniques are available to control the crystal orientation (texture) of films. In this OAD technique, the flux arrives at the substrate with an angle α measured with respect to the surface normal. (When α is large, this technique is called glancing-angle deposition. [35]) Figure 4 shows the effects of physical shadowing during OAD. Islands of different heights are initially nucleated at the surface. Subsequently, the incident flux of material that strikes the surface with an oblique angle α is preferentially deposited onto the top of surface features with larger height values. This preferential growth dynamic gives rise to the formation of well-separated nanostructures. Crystal texture selection also occurs during growth. It is believed that the growth speed competition between crystal planes and the maximization of the capture area facing the flux can lead to the biaxial texture selection [36].

Well-separated nanostructures, even if they possess good biaxial texture quality, are not suitable for use as a buffer layer. For a good buffer layer, one needs a continuous film with no major gaps or voids. Therefore in the past, this technique (sometimes called the incline substrate deposition) has been restricted to a small oblique angle α [7, 8, 11, 12, 37]. Another challenge is that the films normally contain an asymmetric texture orientation with respect to the surface normal due to the oblique-angle-deposition flux. To achieve symmetric pole figure from films grown under large oblique incident angle we developed a flipping rotation mode that is described next.

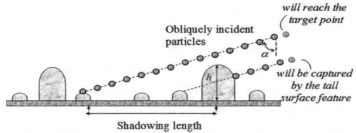

Figure 4. A schematic illustrates the effect of shadowing during oblique-angle deposition. Islands of different height are initially nucleated at the surface. Subsequently, the incident flux of material that strikes the surface with an oblique angle α is preferentially deposited onto the top of surface features that have larger height values.

Flipping rotation OAD setup

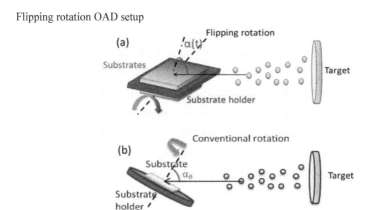

Figure 5. Schematics of experimental setups for film preparation under (a) flipping rotation and (b) conventional rotation. The black dashed line represents the axis of rotation. The axis is in the substrate plane for the flipping rotation, whereas the axis is perpendicular to the substrate plane for the conventional rotation. In the flipping rotation the incident flux angle α(t) from the target relative to the substrate normal (dash-dotted line) is a function of time. In the conventional rotation mode the α_0 is a constant value. ©IOP Publishing, Reproduced by permission of IOP publishing. From [44].

Figures 5(a) shows a schematic of the experimental setup for the flipping rotation mode in oblique angle sputter deposition. The rotational axis is in the substrate plane and parallel to the substrate plane. The rotational axis is also perpendicular to the incident flux direction. This means the angle between the axis of rotation and the direction of incident flux on the substrate is fixed at 90°. The curved arrow represents the rotation direction. Therefore, the incident flux angle $\alpha(t)$ changes as the substrate rotates in a flipping mode. This is in contrast to the conventional setup shown in Fig. 5(b) where the axis of rotation is perpendicular to the substrate plane and the flux incident angle is fixed as the substrate rotates. Two samples can be placed on each side of the substrate holder in the flipping rotation mode whereas only one sample can be placed on the substrate holder in the conventional rotation mode.

Growth conditions for different samples
- Oblique angle vapor deposited biaxial CaF_2 nanostructures on amorphous substrates

CaF_2 films were deposited on native oxide covered Si(100) substrates using the OAD technique and studied extensively as a function of incident angle, substrate temperature and thickness [38, 39]. The deposition conditions were: source-substrate distance 0.25 m, base pressure 2×10^{-8} Torr, average deposition pressure 5×10^{-7} Torr, and the normal deposition rate 13 nm/min. The temperatures were measured at the front surface of the samples.
- Oblique angle vapor deposited Ge film on biaxial buffer CaF_2 film on glass

Corning 2947 glass (Corning Inc., Corning, NY) was used as substrates for the growth of biaxially textured CaF_2 films. The CaF_2 incident flux angle was 65° with respect to the surface normal. This CaF_2 film consisted of ~900 nm tall vertically aligned nanorods. Then this film was capped by ~200 nm thick CaF_2 layer deposited under normal vapor incidence to fill the voids between nanorods. The normal deposition rate for CaF_2 was ~15 nm/min. The details of this deposition procedure have been described elsewhere [40]. The Ge vapor depositions on CaF_2/glass were carried out in a separate high vacuum chamber with a base pressure of 5×10^{-8} Torr. The substrate temperature could be varied and it was determined by a K-type thermocouple placed on the surface of the glass. The substrate to Ge source distance was ~30 cm and the pressure during deposition was ~8×10^{-7} Torr. The deposition rate and film thicknesses of Ge films were monitored by a quartz crystal microbalance (QCM) and were verified through cross section scanning electron microscope (SEM) images. The normal deposition rate for the Ge was ~10 nm/min. Ge films on unbuffered (bare glass) and CaF_2 buffered glass substrates at various substrate temperatures were simultaneously deposited by thermal evaporation. The Ge films on CaF_2 buffered layer possessed the best texture when the substrate temperature was 400 °C.
- MOCVD deposited CdTe film on biaxial buffer CaF_2 on amorphous substrates

The amorphous native oxide covered Si(100) substrates cut from a wafer were used as the substrates. The biaxially textured CaF_2 nanorods of ~700 nm height were grown using oblique angle vapor deposition at 55° incident flux angle and ~18 nm/min deposition rate in a thermal evaporator (described elsewhere [39]). A CdTe film was grown on the CaF_2 nanorods using a vertical, low-pressure, (50 Torr) metal-organic chemical vapor deposition (MOCVD) reactor [41]. After loading the CaF_2 nanorods sample into the reactor, the chamber was pumped down to a pressure of ~1 Torr. The Cd source was dimethylcadmium and was kept at 2 °C. The Te source was diethyltelluride and was kept at 16 °C. Hydrogen was the carrier gas. After the chamber pressure reached about 1 Torr, the growth pressure was slowly brought up to 50 Torr, while the temperature was raised to 425 °C under a hydrogen flow rate of 2.5 slm (standard liter per min). The temperature of the susceptor was allowed to stabilize at 425 °C for 30 minutes before the CdTe film growth. The growth of CdTe film had three stages: nucleation growth layer, annealing, and fast growth layer. First, a nucleation layer of CdTe was grown on the CaF_2

nanorods for 10 minutes at 425 °C. Next, the sample was annealed at a higher temperature (600 °C) for 30 minutes under Te over-pressure. A 2 μm thick CdTe layer was then grown for 110 minutes at 425 °C.

- Flipping rotation and conventional rotation deposited Mo films on amorphous substrates

The Mo nanostructured samples were grown on native oxide covered Si(100) substrates of about 2×2 cm^2 size by rotating the substrates dynamically in a DC magnetron sputtering system. The base pressure in the chamber was $\sim 6.5 \times 10^{-7}$ Torr. The Ar gas flow rate was controlled at 2 sccm, the working pressure of Ar was set to 2.3 mTorr, and the power was set to 200 W. The Mo target with 99.95% purity was a 3" round disk and the distance from the target surface to the center of the substrate was ~15 cm. One film was grown by flipping rotation and another film was grown by conventional rotation. Both films were grown with a 1 RPM (rotations per minute) rotation speed. The incident angle α_o was fixed at 85° for conventional rotation and the incident angle $\alpha(t)$ was varied as a function of time t as the substrate was flipping in the flipping rotation mode.

RESULTS AND DISCUSSION

Morphology and texture of biaxial CaF$_2$ nanorods

Figures 6(a) and 6(b) show the top view and side view SEM images of ~200 nm thick CaF$_2$ nanorods film deposited at $\alpha = 60°$ and room temperature. Figure 6(c) shows the x-ray {111} pole figure. The arrow from right to left in figure 6(a) and the arrow from left to right in Fig. 6(c) are the projections of the incident vapor direction onto the substrate plane. The top view image in Fig. 6(a) shows the slanted facets facing the incident vapor direction. In Fig. 6(b), the incident vapor direction with respect to the substrate normal is represented by a slanted arrow. The CaF$_2$ film is composed of faceted columns and the columns are almost vertical. The porosity of the film increases as the incident flux α increases.

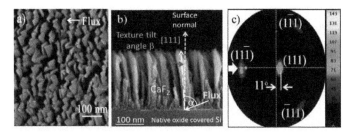

Figure 6. (a) Top view SEM image of a CaF$_2$ film deposited on native oxide covered Si(001) wafer at incident angle $\alpha = 60°$ and room temperature. (b) Side view SEM image of CaF$_2$ nanorods. The red arrow points to the tilt angle β between the texture axis [111] (white long dashed line) and the surface normal (white dashed line). The scale bar is 100 nm. (c) X-ray {111} pole figure of the CaF$_2$ nanorods. The center (111) pole is about 11° away from the substrate normal. The white arrows are the projection of the incident vapor direction in the substrate plane. ©IOP Publishing, Reproduced by permission of IOP publishing. From [38].

The X-ray intensity vs. 2θ spectra (not shown here) of the film shows that the CaF₂ film has a dominant (111) peak and is highly textured. Thus the CaF₂ film has a preferential out-of-plane orientation along the [111] direction. To see if this film is biaxially oriented (having alignments along both the out-of-plane and in-plane directions), we show X-ray pole figure measurements for the (111) diffraction peaks in Fig. 6(c). The [111] direction is not exactly at the center of the {111} pole figure and is tilted by 11±1° away from the incident flux. The CaF₂ pole figure is similar to the theoretical (111) poles projected along the [111] direction of a cubic crystal with the [111] direction tilted ~11° off the substrate normal and away from the incident flux. This texture tilt angle β of ~11±1° is also indicated in Fig. 6(b). The pole figure shows that the CaF₂ film is biaxially textured with some dispersions in the poles. The in-plane dispersion for this film deposited at 60° is on the order of 15±5° where the ±5° error is due to the 5° step size in the azimuthal direction for data acquisition.

Biaxial Ge film grown by vapor deposition on biaxial buffer CaF₂ film on glass

Figures 7(a) shows the side view of the SEM image of the Ge/CaF₂/Glass samples grown at 400 °C substrate temperature [42]. The SEM image shows a ~600 nm thick Ge film on the CaF₂ buffered glass substrate and the Ge film is rough with surface asperities on the order of 50-200 nm in size. The XRD line profiles of the Ge films grown simultaneously on glass and on CaF₂/glass at a substrate temperature of 400 °C were collected (not show here). The spectrum of the Ge film grown on CaF₂/glass is highly oriented with the [111] as the out-of-plane orientation. However, the Ge film grown on glass was polycrystalline (with a slight preference for the [111] texture).

To further confirm the out-of-plane direction in the Ge film and to determine the in-plane orientation of Ge film, x-ray pole figure data were collected and analyzed. Figures 7(b) shows the Ge(111) pole figures of the sample grown at 400 °C. This Ge pole figure with the ~11° texture tilt angle off the substrate normal is also similar to the CaF₂ pole figure shown in Fig. 6(c). Therefore, the out-of-plane orientation of the Ge film is the [111] direction. The dispersion in the [111] out-of-plane orientation of the Ge film was approximately ±3 degrees. The Ge (111) pole figure shows two kinds of poles. One set of poles has a weak intensity, labeled as (11$\bar{1}$), ($\bar{1}$11) and (1$\bar{1}$1) and the other set of poles is rotated by 180° with a strong intensity labeled as

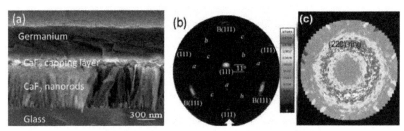

Figure 7. (a) SEM side view image of Ge/CaF₂/glass grown at 400 °C substrate temperature by thermal vapor evaporation. From [41]. (b) X-ray {111} pole figure of the Ge film that rotates 180° about the [111] direction of CaF₂. The central (111) pole is tilted ~11° from the surface normal and away from the incident flux direction (white arrow). The poles labeled a, b, and c are twins about B(1$\bar{1}$1), B(11$\bar{1}$), and B($\bar{1}$11) positions, respectively. From [40]. (c) X-ray {220} pole figure of a Ge film deposited on glass substrate without biaxial CaF₂ layer.

B(11$\bar{1}$), B($\bar{1}$11) and B(1$\bar{1}$1). The Ge poles indicated by B(11$\bar{1}$), B($\bar{1}$11) and B(1$\bar{1}$1) can be either from the growth twins of the type A epitaxial film where the Ge film orientation is same as the CaF$_2$ substrate, or from the type B epitaxial film where the Ge film is rotated 180° about the [111] direction of the CaF$_2$ film. However, the extremely high intensity of the 180° rotated poles suggests that type B epitaxy was preferential in the growth of Ge on CaF$_2$. The pole positions labeled a, b and c, in Fig. 7(b), are the result of twinning about B(1$\bar{1}$1), B(11$\bar{1}$), and B($\bar{1}$11) positions, respectively. In contrast, the CaF$_2$ film did not exhibit any poles corresponding to growth twins, see Fig. 6(c). The twining in Ge film and no twining in CaF$_2$ are also supported by selected area diffraction from Ge and CaF$_2$ layers, respectively, and the data has been presented elsewhere [40].

The type B orientation of Ge(111) on CaF$_2$ buffer layer is further supported by the high resolution TEM image shown in Fig. 8(a). The red dashed line represents the boundary of Ge and CaF$_2$. The schematic in Fig. 8(b) mimic the high resolution TEM image in Fig. 8(a) and indicates that the Ge film is rotated by 180° with respect to the [111] direction of CaF$_2$. Thus the Ge is a type B epitaxy layer.

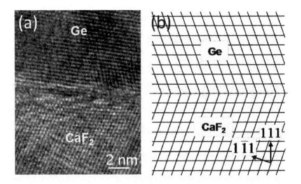

Figure 8. (a) High resolution TEM image of biaxial Ge and CaF$_2$ films. The red dashed line represents an interface of Ge and CaF$_2$. (b) A schematic representing the relative orientation between Ge and CaF$_2$ films. From [42].

First principles calculations using the density function theory has been performed to determine the energetics of the Ge(111) on CaF$_2$ interface. It was found that the type B orientation of the Ge film is favorable among other orientations. The Ge atoms bond to Ca^{2+} and the top F$^-$ layer of CaF$_2$ is removed during the Ge deposition [43].

In contrast to the pole figure of Ge/CaF$_2$/glass, the pole intensity distribution of a Ge film grown directly on a glass substrate without the CaF$_2$ buffer shows a ring structure in the {220} x-ray pole figure. See Fig. 7(c). This indicates a fiber texture with only a preferred out-of-plane orientation but no preferred in-plane orientation. Therefore, it is crucial to have a biaxial layer to induce the epitaxial Ge film.

The angular misorientation between adjacent grains and the grain boundary structure in Ge film can be obtained using high resolution TEM imaging. Figure 9(a) shows a high-resolution TEM image of such a grain boundary (red dashed line) in Ge film. Fig. 9(b) shows a grain misorientation of ~2°. Angular misorientation between adjacent grains observed by TEM is typically much smaller than the angular spread observed in the x-ray pole figures. Adjacent grains misorientation is what dictates the electronic and electrical properties of the film. Because of the very small angle grain boundaries, it is challenging to measure accurately the grain size. From the very small number of grain boundaries observed, we estimated that the grain size should be at least two microns, larger than the thickness of the film (~600 nm).

Figure 9. (a) High resolution TEM image of a grain boundary in the Ge film. (b) A schematic showing a 2° small angle grain boundary. From [40].

Biaxial CdTe film grown by MOCVD on biaxial buffer CaF_2 film on amorphous substrate

Due to the biaxial nature of the CaF_2 layer deposited on the amorphous substrates, the resulting heteroepitaxial semiconducting films are expected to be biaxial. Figure 10(a) shows the cross section SEM micrograph of a 2 micron thick CdTe film deposited by the MOCVD technique onto the CaF_2 buffer layer (with CaF_2 capping layer formed by normal incidence deposition after oblique angle deposition of CaF_2 nanorods) on a glass substrate. Figure 10(b) shows the (220) x-ray pole figure of the CdTe film shown in Fig. 10(a). It is strongly biaxial in that the pole intensity distribution is localized at specific positions labeled in the figure. The six poles in the pole figure are not exactly symmetric with respect to the center of the pole figure. The white cross indicates that the center of the six poles is off the surface normal by about 10° and is away from the incident flux direction.

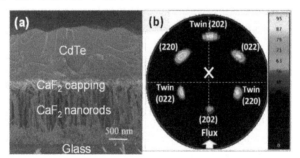

Figure 10. (a) Side view of SEM image of CdTe on CaF_2 nanorods on glass with CaF_2 capping. (b) x-ray {220} pole figure. The white arrow is the projection of incident flux direction on the substrate plane. The white cross is the center of the six poles.

To study the grain boundary in the CdTe film grown on biaxial CaF_2 layer we show the bright field TEM cross sectional image in Fig. 11(a). This particular sample has CaF_2 nanorods and does not have a CaF_2 capping layer. The red dotted curves indicate grain boundaries. The image indicates that the in-plane grain to grain misorientation in the CdTe film to be $\leq 6°$, a small angle boundary. A schematic representation is shown in Fig. 11(b).

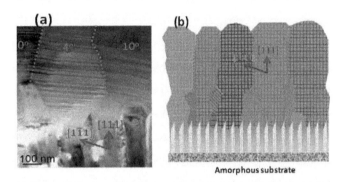

Figure 11. (a) Bright field TEM cross sectional image of CdTe film deposited by MOCVD on CaF_2 nanorods. The grains in the CdTe film have grain boundaries less 6°. (b) A schematic showing the grains and gran boundaries in the CdTe film. From [42].

The interface lattice mismatch between CdTe and CaF_2 is ~20 %. This large lattice mismatch leads to a concentrated array of misfit dislocations at the heteroepitaxial interface. Despite this large lattice mismatch, the high-resolution TEM image (not shown here) indicates every five CdTe lattice matches to six CaF_2 lattice. This 5 to 6 matching at the interface leaves a remaining 1.1 % strain between CdTe and CaF_2. This strain is distributed in the CdTe film and the compliant CaF_2 nanostructured film. The nanoscale CaF_2 rods with gaps in between rods allow additional stress relief through the lateral deformation in the CdTe epilayer. This is in contrast to the CdTe film grown on a continuous CaF_2 capping layer where the stress may mainly be relieved in the epilayer through vertical deformation of the epilayer.

Biaxial Mo and W films grown by flipping rotation OAD

The above biaxial Ge and CdTe semiconductor films were grown on oblique angle deposited CaF_2 buffer layer on amorphous substrates. The biaxial CaF_2 is an insulator and may not be suitable for devices that need electrical conduction. One may ask if it is possible to use a biaxial metal film as a buffer layer to grow epitaxial semiconductors for solar cell and display applications. Recently we developed a flipping rotation method in the dynamic oblique angle deposition technique. This flipping rotation method produces biaxial metallic films such as Mo and W on amorphous substrates [44]. This opens up another route to grow heteroepitaxial semiconductor films directly on metallic surfaces.

We have used the flipping rotation mode to grow Mo and W films on amorphous substrates. Figures 12(a) and (b) show side view and top view SEM images. Figure 12(c) is a reflection high energy electron diffraction (RHEED) pole figure obtained from a Mo film grown by the flipping rotation mode. In contrast, very different results are shown in Figs. 12(d), (e), and (f) from Mo films grown by the conventional rotation. Both films were grown with a 1 RPM (rotations per minute) rotation speed and the incident angle α_o = 85°. Both rotational modes produce vertical nanocolumns of a diameter around 100 nm. However, the density of nanorods differs. For the flipping rotation, the nanorods are more densely packed and the morphology seems to be anisotropic parallel to the rotational axis (vertical dashed line in Fig. 12(b)). The nanorods produced by the conventional rotation shown in Fig. 12(e) are more separated from each other and the morphology does not have an obvious anisotropy. The most striking difference is the texture. Figure 12(c) shows that the RHEED pole figure of Mo grown by the flipping rotation mode has a biaxial texture whereas Fig. 12(f) shows that the RHEED pole figure of Mo grown by the conventional rotation has a fiber texture [44].

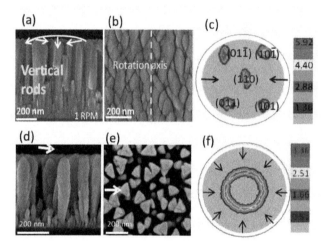

Figure 12. (a) Side view and (b) top view SEM images of 650 nm tall Mo nanorods prepared under the flipping-rotation mode. (d) Side view and (e) top view SEM images of 400 nm tall Mo nanorods prepared under conventional rotation mode. Straight white arrows in (a) represent the instantaneous flux directions during a clockwise substrate flipping rotation. The flux evolution is represented by the curved arrows in (a). The white arrows in (d) and (e) represent the incident flux direction. (c) and (f) are the {110} RHEED surface-pole figures corresponding to films in (a) and (d), respectively. The arrows in (c) show two representative flux directions during rotation, and the flux plane is perpendicular to the pole-figure plane. The arrows in (f) show the flux directions under the conventional rotation mode. The vertical color bar indicates the strength of the diffraction intensity. From [44].

CONCLUSION

We have shown that large oblique angle ($\geq 60°$) vapor deposited CaF_2 nanorods on amorphous substrates including glass has a biaxial texture {111}<121> with no twins. The out-of-plane orientation is [111] and the in-plane orientation is [121]. This biaxial texture preserves even after a normal deposition to form a more continuous CaF_2 capping layer. Despite the large lattice mismatch, this biaxial CaF_2 layer has been used as a buffer layer to grow epitaxial Ge films by vapor deposition and CdTe films by MOCVD at 400 °C and 425 °C, respectively. These epitaxial temperatures are lower than temperature obtained by many other deposition techniques. The biaxial Ge(111) film has a nearly single crystal structure with dominant type B twins and with small angle ($\leq 2°$) grain boundaries. The biaxial CdTe film follows the biaxial texture {111}<121> of CaF_2 with a high density of twins. The misorientation angle among grain boundaries is $\leq 6°$. We also developed a new flipping rotation method in dynamic oblique angle deposition that produces biaxial Mo and W metal films. These biaxial metallic films can serve as buffer layer in addition to CaF_2 nanorods to grow semiconductor films.

ACKNOWLEDGEMENT

This work is supported by the NSF DMR. We thank following former and current collaborators. These include P. Snow, T.C. Parker, W. Yuan, N. LiCausi, M. Riley, S. Rao, Drs. C. Gaire, H.-F. Li, S. Lee, T.L. Chan and S.B. Zhang.

REFERENCES

[1]   A. Goetzberger, C. Hebling, and H.-W. Schock, Photovoltaic materials, history, status and outlook, Materials Science and Engineering: R: Reports **40**, 1-46 (2003).

[2]   C. S. Ferekides, U. Balasubramanian, R. Mamazza, V. Viswanathan, H. Zhao, and D. L. Morel, CdTe thin film solar cells: Device and technology issues, Solar Energy 77, 823 (2004).

[3]   E. Bauer, in *Single-Crystal Films Intern. Conf.*, edited by M. H. Francombe and H. Sato (Oxford: Pergamon Press, Pennsylvania, , 1963), p. 43.

[4]   F. Tang, T. Parker, G.-C. Wang, and T.-M. Lu, Surface texture evolution of polycrystalline and nanostructured films: RHEED surface pole figure analysis, Journal of Physics D: Applied Physics **40**, R427 (2007).

[5]   R. T. Brewer and H. A. Atwater, Rapid biaxial texture development during nucleation of MgO thin films during ion beam-assisted deposition, Appl. Phys. Lett. **80**, 3388 (2002).

[6]   M. P. Paranthaman and T. Izumi, High-performance YBCO-coated superconductor wires, MRS Bulletin **August** 533 (2004).

[7]   A. T. Findikoglu, W. Choi, V. Matias, T. G. Holesinger, Q. X. Jia, and D. E. Peterson, Well-oriented silicon thin films with high carrier mobility in polycrystalline substrates, Adv. Mater. **17**, 1527 (2005).

[8]   C. W. Teplin, D. S. Ginley, and H. M. Branz, A new approach to thin film crystal silicon on glass: Biaxial-textured silicon on foreign template layers, J. of Non-Crystalline Solids **352**, 984 (2006).

9   J. Chen and T. Sekiguchi, Carrier recombination activity and structural properties of small-angle grain boundaries in multicrystalline silicon, The Japan Society of Applied Physics **46**, 6489 (2007).

10 W. Choi, V. Matias, J.-K. Lee, and A. T. Findikoglu, Dependence of carrier mobility on grain mosaic spread in <001> oriented Si films grown on polycrystalline substrates, Appl. Phys. Lett. **87**, 152104 (2005).

11 M. P. Chudzik, R. E. Koritala, L. P. Luo, D. J. Miller, U. Balachandran, and C. R. Kannewurf, Mechanism and processing dependence of biaxial texture development in magnesium oxide thin films grown by inclined-substrate deposition, IEEE Trans. Appl. Superconductivity **11**, 3469 (2001).

12 Y. Xu, C. H. Lei, B. Ma, H. Evans, H. Efstathiadis, M. Rane, M. Massey, U. Balachandran, and R. Bhattacharya, Growth of textured MgO through e-beam evaporation and inclined substrate deposition, Supercond. Sci. Technol. **19**, 835 (2006).

13 B. Ma, M. Li, R. E. Koritala, B. L. Fisher, A. R. Markowitz, R. A. Erck, S. E. Dorris, D. J. Miller, and U. Balachandran, Pulsed laser deposition of YBCO films on ISD MgO buffered metal tapes, IEEE Trans. Appl. Superconduct. **13**, 2695 (2003).

14 A. Goyal, Semiconductor-based, Large-Area, Flexible Electronic Devices, US Patent 7906229, March 15 (2011).

15 A. Goyal., {100}<100> or 45 degrees-rotated {100}<100>, semiconductor-based, large-area, flexible, electronic devices, US Patent 8178221, May 15 (2012).

16 C. W. Teplin, M. P. Paranthaman, T. R. Fanning, K. Alberi, L. Heatherly, S.-H. Wee, K. Kim, F. A. List, J. Pineau, J. Bornstein, K. Bowers, D. F. Lee, C. Cantoni, S. Hane, P. Schroeter, D. L. Young, E. Iwaniczko, K. M. Jones, and H. M. Branz, Heteroepitaxial film crystal silicon on Al2O3: new route to inexpensive crystal silicon photovoltaics, Energy Environ. Sci. **4**, 3346 (2011).

17 S. H. Wee, C. Cantoni, T. R. Fanning, C. W. Teplin, D. F. Bogorin, J. Bornstein, K. Bowers, P. Schroeter, F. Hasoon, H. M. Branz, M. P. Paranthaman, and A. Goyal, Heteroepitaxial film silicon solar cell grown on Ni-W Foils, Energy Environ. Sci. **5**, 6052 (2012).

18 C. Gaire, J. Palazzo, I. Bhat, A. Goyal, G.-C. Wang, and T.-M. Lu, Low temperature epitaxial growth of Ge on CaF2 buffered cube-textured Ni, J. Cryst. Growth **343**, 33 (2012).

19 C. Gaire, S. Rao, M. Riley, L. Chen, A. Goyal, S. Lee, I. Bhat, T.-M. Lu, and G.-C. Wang, Epitaxial growth of CdTe thin film on cube-textured Ni by metal-organic chemical vapor deposition, Thin Solid Films **520**, 1862 (2012).

20 V. Selvamanickam, S. Sambandam, A. Sundaram, S. Lee, A. Rar, X. Xiong, A. Alemu, C. Boney, and A. Freundlich, Germanium films with strong in-plane and out-of-plane texture on flexible, randomly textured metal substrates, J. Cryst. Growth **311**, 4553 (2009).

21 J. R. Groves, J. B. Li, B. M. Clemens, V. LaSalvia, F. Hasoon, H. M. Branz, and C. W. Teplin, Biaxially-textured photovoltaic film crystal silicon on ion beam assisted deposition CaF2 seed layers on glass, Energy Environ. Sci. **5**, 6905 (2012).

22 L. J. Schowalter and R. W. Fathauer, Molecular beam epitaxy growth and applications of epitaxial fluoride films, J. Vac. Sci. Technol. A **4**, 1026 (1986).

23 J. Wollschlager, C. Deiter, M. Bierkandt, A. Gerdes, M. Baumer, C. R. Wang, B. H. Muller, and K. R. Hofmann, Homogeneous Si films on CaF2/Si(1 1 1) due to boron enhanced solid phase epitaxy, Surface Science **600**, 3637 (2006).

24 D. Y. Kim, B. J. Ahn, S. I. Moon, C. Y. Won, and J. Yi, Low temperature microcrystalline-Si film growth using a CaF2 seed layer, Solar Energy Materials and Solar Cells **70**, 415 (2002).

25 M. A. Olmstead, Thin Films: Heteroepitaxial Systems, World Scientific, Singapore (1999).

[26] C. R. Wang, B. H. Muller, E. Bugiel, and K. R. Hofmann, Surfactant enhanced growth of thin Si films on $CaF_2/Si(1\ 1\ 1)$, Applied Surface Science **211**, 203–208 (2003).

[27] K. Koike, T. Komuro, K. Ogata, S. Sasaa, M. Inoue, and M. Yano, CaF_2 growth as a buffer layer of ZnO/Si heteroepitaxy, Physica E **21**, 679 (2004).

[28] W. Li, T. Anan, and L. J. Schowalter, Nucleation of GaAs on $CaF_2/Si(111)$ substrates, Appl. Phys. Lett. **65**, 595 (1994).

[29] M. Kessler, A. N. Twari, S. Blunier, and H. Zogg, in *Conference Record of the Twenty Fourth IEEE Photovoltaic Specialists Conference* Hawaii, 1994), p. 323.

[30] H. Zogg, J. Masek, C. Maissen, S. Blunier, and H. Weibel, IV-VI compounds on fluoride/silicon heterostructures and IR devices, Thin Solid Films **184**, 247 (1990).

[31] G. Breton, M. Nouaoura, C. Gautier, M. Cambon, S. Charar, M. Averous, and V. Ribes, Annealing under vacuum and Se flux of CaF_2 molecular beam epitaxy surfaces prior to $PbSe/CaF_2/Si$ growth, Surface Science **406**, 63 (1998).

[32] V. Mathet, P. Galtier, F. Nguyen-Van-Dau, G. Padeletti, and J. Olivier, A microstructural study of crystalline defects in $PbSe/BaF_2/CaF_2$ on (111) Si grown by molecular beam epitaxy, Journal of Crystal Growth **132**, 241 (1993).

[33] F. Tang, G.-C. Wang, and T.-M. Lu, Surface pole figures by reflection high-energy electron diffraction, Appl. Phys. Lett. **89**, 241903 (2006).

[34] B. D. Culllity, *Elements of X-ray Diffraction* (Addison-Wesley, 1978).

[35] K. Robbie and M. J. Brett, Sculptured thin films and glancing angle deposition: growth mechanics and applications, J. of Vacuum Science & Technology A **15 (3)**, 1460 (1997).

[36] S. Mahieu, P. Ghekiere, D. Depla, and R. D. Gryse, Biaxial alignment in sputter deposited thin films, Thin Solid Films **515**, 1229 (2006).

[37] P. Ghekiere, S. Mahieu, G. D. Winter, R. D. Gryse, and D. Depla, Scanning electron microscopy study of the growth mechanism of biaxially aligned magnesium oxide layers grown by unbalanced magnetron sputtering, Thin Solid Films **493**, 129 (2005).

[38] C. Gaire, P. Snow, T.-L. Chan, M. Yuan, M. Riley, Y. Liu, S. B. Zhang, G.-C. Wang, and T.-M. Lu, Morphology and texture evolution of nanostructured CaF_2 films on amorphous substrates under oblique incidence flux, Nanotechnology **21**, 445701 (2010).

[39] H.-F. Li, T. Parker, F. Tang, G.-C. Wang, and T.-M. Lu, Biaxially orientated CaF_2 films on amorphous substrates, J. Cryst. Growth **310**, 3610 (2008).

[40] C. Gaire, P. C. Clemmer, H.-F. Li, T. C. Parker, P. Snow, I. Bhat, S. Lee, G.-C. Wang, and T.-M. Lu, Small angle grain boundary Ge films on biaxial CaF_2/glass substrate, J. Cryst. Growth **312**, 607 (2010).

[41] W. Yuan, F. Tang, H.-F. Li, T. Parker, N. LiCausi, T.-M. Lu, I. Bhat, G.-C. Wang, and S. Lee, Biaxial $CdTe/CaF_2$ films growth on amorphous surface, Thin Solid Films **517**, 6623 (2009).

[42] T.-M. Lu, H. Li, C. Gaire, N. Licausi, T.-L. Chan, I. Bhat, S. B. Zhang, and G.-C. Wang, Quasi-single Crystal Semiconductors on Glass Substrates through Biaxially OrientedBuffer Layers, Mater. Res. Soc. Symp. Proc. **1268**, EE03 (2010).

[43] T.-L. Chan, C. Gaire, T.-M. Lu, G.-C. Wang, and S. B. Zhang, Type B epitaxy of Ge on $CaF_2(111)$ surface, Surface Science **604**, 1645 (2010).

[44] L. Chen, T.-M. Lu, and G.-C. Wang, Biaxially textured Mo films with diverse morphologies by substrate-flipping rotation, Nanotechnology **22**, 505701 (2011).

PHOTOINDUCED HYDROPHILICITY AND PHOTOCATALYTIC PROPERTIES OF Nb_2O_5 THIN FILMS

Raquel Fiz, Linus Appel and Sanjay Mathur*
Institute of Inorganic Chemistry, University of Cologne
Greinstr. 6, 50939 Cologne, Germany

ABSTRACT

Self-cleaning properties can be achieved by designing nanostructured superhydrophobic surfaces that repel dirt or alternatively by superhydrophilic coatings that can decompose organic matter through photocatalytic processes. While the photocatalytic properties and the related photoinduced hydrophilicity of TiO_2 have been widely investigated, niobium pentoxide (Nb_2O_5) remains as a less investigated but promising material for transparent coatings with self-cleaning properties. In this work, Nb_2O_5 thin films were fabricated by the decomposition of a molecular precursor $[Nb(O^iPr)_5]_2$ in a low pressure chemical vapor deposition (LPCVD) reactor at different substrate temperatures (500-1000 °C). Larger crystallites with preferred needle-like textures were observed in Nb_2O_5 deposits obtained at higher temperatures (> 800 °C), whereas smoother films were observed in the lower temperature regime suggesting the change in the balance of forces and growth kinetics. We report here on the wetting and photocatalytic properties of CVD-grown Nb_2O_5 thin films and correlate their functional properties with the surface morphology in order to demonstrate the potential of Nb_2O_5 coatings in smart window and self-cleaning applications.

INTRODUCTION

The natural self-cleaning effect observed on the lotus (*Nelumbo nucifeara*) and other plants has inspired the development of superhydrophobic surfaces for anti-fogging, anti-bacterial, self-cleaning, and drug-delivery applications[1]. The correlation between microstructure and wettability demonstrated by Barthlott and Neinhuis as the Lotus effect[2] had triggered significant interest in the fabrication of hierarchical nanostructures to enhance the functionality of inorganic surfaces. Surface wettability is an important property relevant to solid-state materials and it is commonly quantified by measuring the contact angle (CA) at which a liquid-vapor interface contacts a solid-liquid interface. Hydrophobic surfaces exhibiting high contact angles (> 90°) are highly desired for easy-to-clean applications due to surface texture. Alternatively, the self-cleaning effect can be induced by surface coatings based on photocatalytic materials.

Superhydrophilic surfaces (CA < 10 °) of semiconductor oxides such as titania (TiO_2) are known to exhibit self-cleaning properties due to their photocatalytic behaviour. The photogenerated holes and electrons at the surface of a semiconductor material involve oxidation and reduction processes that result in the decomposition of most of the organic compounds due to the formation or highly reactive radicals at the semiconductor surface, which are subsequently washed out by (rain) water due to the low contact angles of the surface. In particular, titanium dioxide (TiO_2) has attracted extensive attention in solar cell and photocatalysis applications due to its chemical inertness, high photostability, nontoxicity, low cost fabrication and high photoreactivity[3]. In addition to the conventional TiO_2 photocatalysis, Hashimoto et al. reported the self-cleaning and antifogging properties of this material based on the photoinduced hydrophilicity properties[4,5]. Therefore, surface wettability coupled with surface morphology plays an important role in the design of smart materials for self-cleaning applications that are able to switch between hydrophobic and hydrophobic properties in response to an external stimulus.

Niobium pentoxide (Nb_2O_5) is an n-type semiconductor material with band gap (E_g=3.4 eV) and electron injection efficiency comparable to TiO_2 that makes it a promising material for self-cleaning and optoelectronic applications[6], especially due to its superior chemical stability in air and in water as well as corrosion resistance in both acid and basic media. Due to its acidic character, ascribed to the presence of cations in high valence states, one of the major applications of Nb_2O_5 is in heterogeneous catalysis[7], whereby niobia compounds have been used as active phase, support, solid acid catalyst or redox material[8,9,10,]. Furthermore, its higher conduction band edge (see figure 1) slightly above that of TiO_2 promises a more efficient photocatalytic activity[11]. Nevertheless, the properties of Nb_2O_5 strongly depend on the synthetic method used[6] and among the reported methods chemical vapour deposition (CVD) allows high performance coatings on large-scale substrates with a high degree of structural control and coverage homogeneity and represents one of the preferred methods to fabricate functional coatings and other nanostructures on glass, ceramics, semiconductor or plastics substrates[12,13,14,15,16].

EXPERIMENTAL

$[Nb(O^iPr)_5]_2$ synthesized according to the procedure reported by Bradley et al.[17] was used as a single-source precursor in a horizontal cold-wall chemical vapor deposition (CVD) reactor[12]. Films were grown at different substrate temperatures (500-1000 °C) on silicon and quartz substrates, which were previously cleaned in a solution of ethanol and isopropanol 2:1 through an ultrasonic bath treatment (15 min).

The structural properties of the film were investigated by atomic force microscopy (AFM) performed on a XE-100 Park system equipped with 910 ACTA cantilever, operating in the tapping mode. The results were analysed in XEI software, from which the root mean square roughness (R_{RMS}) value and average grain size were estimated.

The contact angles (CAs) were measured at room temperature on a DSA 100 Kruess instrument using the sessile drop fitting method and deionized water (5 µL) as the test solution. CAs were determined using the circular method. Each sample was measured with 3 drops and 10 images were taken for each droplet. The average value with the corresponding deviation was chosen as the representative image.

Photocatalytic activity of the samples was tested by the decomposition a methylene blue (MB) solution with an initial concentration of $3^x 10^{-8}$ M. Nb_2O_5 coated silicon substrates (001) were immersed in the solution, and irradiated by a 150W Xenon arc lamp (Oriel). The absorbance of the methylene blue (MB) solution was periodically measured by a UV-spectrometer (Lambda 950 UV/vis spectrometer from Perkin Elmer in transmission mode). The decay in absorption corresponded to a variation in concentration, which was estimated by the Beer-Lambert's law. The semilog plot of the decay of the absorption intensity at 665 nm $vs.$ time was fitted to a straight line. The linearity is consistent with the pseudo first-order reaction and the rate constant could be estimated from the linear regression. The surface areas of substrates were measured using digital photographs of the substrates, which was kept at 0.25 cm^2 in all the cases.

RESULTS AND DISCUSSION

The surface acidity, redox properties and photocatalytic activities of a material are strongly linked to its intrinsic structural properties and surface features. There are few reports on the photocatalytic activity of Nb_2O_5[11,18,19,20], however, to the best of our knowledge, the photocatalytic behaviour and photoinduced hydrophilicity properties of Nb_2O_5 thin films deposited by CVD, and their correlation to surface morphology has not been so far investigated. In this work, we report on the influence of the surface roughness and grain size of Nb_2O_5 thin films on their photocatalytic and wetting properties.

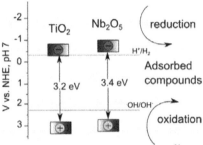

Figure 1. Energy band diagram of TiO_2 and Nb_2O_5 at pH=7[21]. The redox potential positions of H^+/H^2 and OH/OH^- are also illustrated as reference.

The surface topography of the Nb_2O_5 thin films deposited on Si (001) substrates at different substrate temperatures was imaged by atomic force microscopy in tapping mode (see figure 2), which confirmed temperature-dependent grain growth and evolution. Based on the AFM measurements, low process temperatures (500 °C) result in the formation of small Nb_2O_5 nanocrystals of an average size of 15 nm, whereas substantially larger grain sizes (50 nm) were observed at higher (1000 °C) temperatures (Table 1). AFM images of Nb_2O_5 deposited at high temperatures (Figure 2 (d-f)) reveal a columnar growth possibly associated with the Volmer-Webber growth mechanism, in which interactions among adtoms are stronger than those between adatom and the substrate surface, leading to the formation of clusters growing with preferential directions. that the finding that maximum root mean square roughness of the Nb_2O_5 thin film deposited at 900 °C is higher than the one obtained for the film deposited at 1000 °C, what can be attributed to the AFM measurement itself, in which the tip of the cantilever is not completely able to follow the morphology of the grains due to the vertical orientation of the grains.

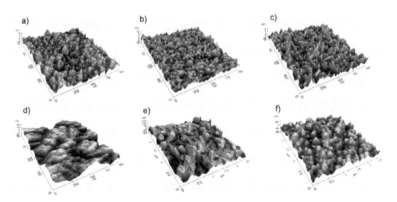

Figure 2. AFM images of Nb_2O_5 thin films deposited on silicon substrates at (a) 500 °C, b) 600 °C, c) 700 °C, d) 800 °C, e) 900 °C, f) 1000 °C.

The wettability of the as-prepared Nb_2O_5 films was evaluated by measuring the contact angle (CA) of water droplets at room temperature under air. As-deposited samples exhibited contact angles varying from 66.7 ± 0.1 ° to 105.6 ± 0.2 °. As expected, we observe a correlation between wettability and microstructure of the films: the hydrophobic surface (CA= 105.6 ± 0.16 °) is ascribed to the needle-like morphology of the film. According to Wenzel's equation, the contact angle of a rough surface is different from the intrinsic contact angle for a certain material and, therefore, the surface microstructure plays a decisive role in the hydrophobicity of the layer[22].

Table 1. Contact Angle (CA) of the Nb_2O_5 nanostructured films before and after UV Irradiation

Substrate Temperature* [°C]	Average Grain Size [nm]	Roughness (R_{RSM}) [nm]	Contact Angle (CA) after CVD process [°]	Contact angle (CA) after 90 min UV Irradiation [°]
500	15± 0.6	2.5	66.7± 0.1	5.1± 0.1
600	18± 1.4	2.9	67.0± 0.2	4.8± 0.1
700	24± 5.0	7.0	67.9± 0.1	4.5± 0.0
800	47± 2.6	7.2	69.6± 0.0	4.2± 0.0
900	45± 0.6	40.8	70.4± 0.1	4.1± 0.1
1000	50± 1.4	35.4	105.6± 0.2	4.2± 0.1

*Substrate temperature used in the process under the decomposition of $Nb_2(O^iPr)_{10}$. CA values were dependent mainly on the morphology and not on the precursor used.

The wetting properties of the samples could be reversibly tuned from hydrophobic to hydrophilic upon UV irradiation for 90 min at 256 nm. The contact angles of the films decreased upon UV exposure to 4.1 ± 0.1 °- 5.1 ± 0.1 ° ascribable to superhydrophilicity caused by surface redox reactions reported for Nb_2O_5[18]. UV irradiation produces oxygen vacancies at the surface of the Nb_2O_5 surface due to the reaction of photogenerated holes and lattice oxygen. Water molecules can be adsorbed onto the surface oxygen vacancies (hydroxyl formation) generating a hydrophilic surface. It is noticeable that the contact angle of the Nb_2O_5 films deposited at different temperatures shows similar values after UV irradiation, indicating that the hydrophilic properties are primarily due to the changes in the surface chemistry and that the surface morphology plays only a minor role. Figure 3 shows the change in the water contact angle of Nb_2O_5 needle-like film deposited at high temperature as a function of UV light irradiation time.

The initial contact angles of the Nb_2O_5 films could be recovered to original values upon storing the samples for 48 hours in the dark. The surface is slowly oxidized by atmospheric oxygen and, consequently, the hydroxyl groups are cleared due to a decreasing number of photogenerated holes at the surface. The hydrophobic properties of the Nb_2O_5 nanorods were also recovered as apparently the water filled in the voids created by vertically grown grains is gradually moved up and evaporated. Such recovery times and reversible wetting behavior can be advantageous in self-cleaning applications.

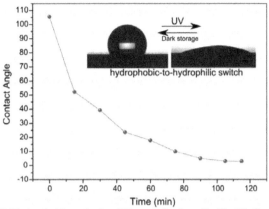

Figure 3. Hydrophobic-to-hydrophilic switch of needle-like Nb_2O_5 thin films.

Besides the photoinduced superhydrophilicity, the UV irradiation of Nb_2O_5 thin films is related to the photocatalytic properties. The photocatalytic activity of the samples was evaluated by decomposition of methylene blue (MB) under UV illumination. The photogenerated electrons (conduction band) and holes (valence band) are involved in redox reactions that lead to a degradation of the dye in aqueous solution. MB absorbs light of wavelengths of 612 and 665 nm, respectively, which allowed following the photocatalytic properties by absorption spectroscopy measurements (Figure 4). The photodecomposition of MB on Nb_2O_5 films as a function of the irradiation time (by selecting the decrease in the absorption intensity value at 665 nm) showed a significant improvement of photocatalytic properties based on the decay constants and decomposition rates, for films composed of needle-like Nb_2O_5 deposits. Besides changes in morphology, the deposition temperature is expected to influence the phase and crystallinity of the deposits due to the polymorphic nature of Nb_2O_5: Brauer distinguished in 1941 three modifications of Nb_2O_5 corresponding to the synthesis temperature at low (T), medium (M) and high (H) temperatures[23] and further modifications have been reported since then[24]. Although investigations on the phase and defect structure of Nb_2O_5 should be further investigated, the higher reactivity observed in Figure 4 is attributed to an enhanced surface area that results in a more effective photocatalytic activity. Moreover, the surface becomes more hydrophilic upon UV irradiation, which explains the faster hydrophobic-to-hydrophilic switching interesting for practical applications. The curves suggest a pseudo first-order reaction, in which the overall rate constants were determined from the slopes of linear regression analysis (Table 2).

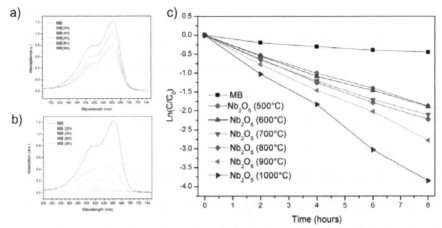

Figure 4. a) Decomposition of methylene blue in aqueous solution by UV irradiation b) Decomposition of methylene blue solution in presence of a needle-like Nb_2O_5 thin film coated on a silicon substrate. c) Logaritmic plot of the concentration change of the methylene blue solution vs. UV-irradiation time of Nb_2O_5 thin films deposited at different substrate temperatures in the CVD process.

Table 2. Decay constants of the methylene blue absorption in Nb_2O_5 films deposited by decomposition of $[Nb_2(O^tPr)_{10}]$.

Substrate Temperature [°C]	Decay constant [h^{-1}]*	Linear correlation coefficient [R^2]	Normalized decay constant [h^{-1}cm^{-2}] [†]
Nb_2O_5 500	0.23± 0.01[‡]	0.997	0.92± 0.03[‡]
Nb_2O_5 600	0.24± 0.01[‡]	0.992	0.94± 0.04[‡]
Nb_2O_5 700	0.26± 0.01[‡]	0.988	1.05± 0.06[‡]
Nb_2O_5 800	0.28± 0.01[‡]	0.993	1.12± 0.05[‡]
Nb_2O_5 900	0.34± 0.01[‡]	0.997	1.36± 0.04[‡]
Nb_2O_5 1000	0.49± 0.02[‡]	0.996	1.94± 0.06[‡]

*Absolute value
[†]Nb_2O_5 films were deposited on Si substrates with an area of $0.25cm^2$
[‡]Standard error from the lineal regression fitting.

CONCLUSION

In summary, we have investigated the photocatalytic properties of Nb_2O_5 thin films deposited by chemical vapor deposition, and have correlated their microstructure against the wetting behavior. Contact angle measurements reveal that while hydrophobic properties (CA=105.6± 0.2) attributed to high roughness values are reached for needle-like Nb_2O_5 deposits, a reversible intrinsic photoinduced photohydrophilicity is observed. Furthermore, the irradiation of Nb_2O_5 with UV light influenced the photocatalytic degradation of methylene blue solution, which is comparable to the well-studied TiO_2. Interestingly, a superior photocatalytic efficiency was observed in the needle-like Nb_2O_5 deposits that was attributed to an enhanced surface area. The switchable hydrophobic-to-hydrophilic behavior in combination with the exhibited photocatalytic activity renders the transparent Nb_2O_5 coatings as potential candidates in self-cleaning applications.

ACKNOWLEDGMENT

Authors are thankful to the University of Cologne (Germany Authors are thankful to the University of Cologne and the BMBM initiative LIB-2015 (ProjectKoLIWin) for providing the financial support.

REFERENCES
(1)   Fujishima, A.; Rao, T. N.; Tryk, D. A. *Journal of Photochemistry and Photobiology C: Photochemistry Reviews* 2000, *1*, 1–21.
(2)   Barthlott, W.; Neinhuis, C. *Planta* 1997, 1–8.
(3)   Hashimoto, K.; Irie, H.; Fujishima, A. *Japanese Journal of Applied Physics* 2005, *44*, 8269–8285.
(4)   Wang, R.; Hashimoto, K. *Nature* 1997, *388*, 431–432.
(5)   Wang, J.; Mao, B.; Gole, J. L.; Burda, C. *Nanoscale* 2010, *2*, 2257–61.
(6)   Aegerter, M. a. *Solar Energy Materials and Solar Cells* 2001, *68*, 401–422.
(7)   Nowak, I.; Ziolek, M. *Chem.Rev.* 1999, *99*, 3603–3624.
(8)   Nakajima, K.; Fukui, T.; Kato, H.; Kitano, M.; Kondo, J. N.; Hayashi, S.; Hara, M. *Chemistry of Materials* 2010, *22*, 3332–3339.
(9)   Marzo, M.; Gervasini, A.; Carniti, P. *Catalysis Today* 2012, *192*, 89–95.
(10) Tanabe, K.; Okazaki, S. *Applied Catalysis A: General* 1995, *133*, 191–218.
(11) Prado, A. G. S.; Bolzon, L. B.; Pedroso, C. P.; Moura, A. O.; Costa, L. L. *Applied Catalysis B: Environmental* 2008, *82*, 219–224.
(12) Appel, L.; Fiz, R.; Tyrra, W.; Mathur, S. *Dalton transactions* 2012, *41*, 1981–90.
(13) Mathur, S.; Sivakov, V.; Shen, H.; Barth, S.; Cavelius, C.; Nilsson, a.; Kuhn, P. *Thin Solid Films* 2006, *502*, 88–93.
(14) Mathur, S.; Ruegamer, T.; Grobelsek, I. *Chemical Vapor Deposition* 2007, *13*, 42–47.
(15) Mathur, S.; Barth, S. *Small* 2007, *3*, 2070–5.
(16) Mathur, S.; Barth, S.; Werner, U.; Hernandez-Ramirez, F.; Romano-Rodriguez, A. *Advanced Materials* 2008, *20*, 1550–1554.
(17) Bradley, D. C.; Chakravarti, B. N.; Wardlaw, W. *J. Chem. Soc.* 1956, 2381–2384.
(18) O'Neill, S. a.; Parkin, I. P.; Clark, R. J. H.; Mills, A.; Elliott, N. *Journal of Materials Chemistry* 2003, *13*, 2952.
(19) Esteves, A.; Oliveira, L. C. A.; Ramalho, T. C.; Goncalves, M.; Anastacio, A. S.; Carvalho, H. W. P. *Catalysis Communications* 2008, *10*, 330–332.
(20) Zhao, Y.; Eley, C.; Hu, J.; Foord, J. S.; Ye, L.; He, H.; Tsang, S. C. E. *Angewandte Chemie* 2012, *51*, 3846–9.
(21) Miyauchi, M.; Nakajima, A.; Watanabe, T.; Hashimoto, K. *Chem. Mater.* 2002, *14*, 2812–2816.
(22) Wenzel, R. . *Ind.Eng.Chem.* 1936, *28*, 988–994.
(23) Schäfer, H.; Gruehn, R.; Schulte, F. *Angew. Chem. Internat. Edit.* 1966, *5*, 40–52.
(24) Marucco, J. F. *The Journal of Chemical Physics* 1979, *70*, 649.

HARD NANOCOMPOSITE COATINGS: THERMAL STABILITY, PROTECTION OF SUBSTRATE AGAINST OXIDATION, TOUGHNESS AND RESISTANCE TO CRACKING

J. Musil
Department of Physics, University of West Bohemia, Univerzitní 22, 306 14 Plzeň,
Czech Republic
musil@kfy.zcu.cz

ABSTRACT

The paper briefly reports on the present state of art in the field of hard nanocomposite coatings. It is divided in four parts. The first part is devoted to the enhanced hardness of nanocomposite coatings. The second part is devoted to the thermal stability of nanocomposite coatings, thermal cycling of nanocomposite coatings and formation of amorphous coatings with thermal stability and oxidation resistance above 1000°C using sputtering. The third part reports on new advanced hard nanocomposite coatings with enhanced toughness. The fourth part reports on flexible hard nanocomposite coatings resistant to cracking. It is shown that the hard coatings with enhanced toughness and resistance to cracking represent a new class of advanced protective and functional coatings with a huge application potential. At the end, trends of next development of the advanced hard nanocomposite coatings are outlined.

1. ENHANCED HARDNESS OF NANOCOMPOSITE COATINGS

Nanocomposite coatings represent a new generation of materials [1-2 and references therein]. They are composed of at least two separated phases with nanocrystalline (nc-) and/or amorphous (a-) structure or their combinations. The nanocomposite materials, due to very small ($\leq$10 nm) grains and a significant role of boundary regions surrounding individual grains, exhibit enhanced or even completely new unique properties compared with the conventional materials composed of larger ($\geq$100 nm) grains. This change in behaviour of nanocomposite materials is caused by the increase of the ratio S/V of the grain surface S and its volume V (S/V>0.1). It results in increasing dominance of the grain boundary regions, reduced action of the grain volume, stopping of the generation of dislocations and the promotion of new processes such as the grain boundary sliding or the grain boundary enhancement due to an inter-atomic interaction between the atoms of neighbouring grains with decreasing grain size d. The dramatic change in the behaviour of the nanocomposite materials compared with that of the conventional materials is a result of (1) *the strong change of the geometrical structure of material*, particularly the size d and shape of the grain and the separation distance w between grains, (2) the elemental composition of grains, the crystallographic orientation of grains and the phase composition of the nanocomposite, and (3) the enhanced chemical and electronic bonding between atoms of neighbouring phases or grains. These are main reasons why the nanocomposite coatings can exhibit enhanced hardness and further new unique properties.

The enhanced hardness of the coating can be due to (i) the macrostress generated in the coating during its growth, (ii) the coating nano-structure and (iii) short covalent bonds between atoms. The enhanced hardness of the nanocomposite coating based on its nanostructure strongly depends on the size and shape of nanograins, and the content and X-ray structure of individual phases, see Fig.1. From this figure it is seen that there is a critical value of the grain size $d_c \approx 10$ nm at which a maximum value of hardness H_{max} of the coating is achieved. The region around H_{max}, i.e. around $d=d_c$, corresponds to a continuous transition from the activity of *the intragranular deformation processes* at $d > d_c$, dominated by the dislocations and described by the Hall-Petch law, to the activity of *the intergranular deformation processes* at $d<d_c$ dominated by interactions between atoms of neighbouring grains and/or by the small-scale sliding in grain boundaries. In materials composed of small grains with $d \leq d_c$ the ratio $S/V \geq 0.1$ and the dislocations are not generated. Therefore, the nanostructure of the material starts to play a dominant role.

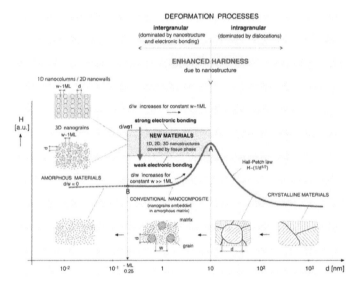

Fig.1. Hardness H of material as a function of size d and shape of grains [1].

2. THERMAL STABILITY OF HARD NANOCOMPOSITE COATINGS

Unique properties of the nanocomposite coatings are determined by their nanostructure. The nanostructure is, however, a metastable phase. It means that in the case when the temperature T under which the coating is operated achieves or exceeds the temperature $T_{nc\ stab}$, corresponding to the thermal stability of the nanostructure of material, the coating starts to be converted to the conventional crystalline material due to the growth of grains. The grain growth results in the destruction of the coating nanostructure and in the formation of new crystalline phases. This is the reason why the nanocomposite coating loses its unique properties at temperatures $T > T_{nc\ stab}$. Therefore, the thermal stability of the nanocomposite coating is usually determined by the stability of its properties with increasing temperature T. This fact is illustrated on the thermal stability of the structure of 7000 nm thick nc-(t-ZrO_2)/a-SiO_2 nanocomposite coatings with a low (~3 at.%) of Zr content exposed to the thermal cycling from room temperature (RT) to a maximal annealing temperature $T_{a\ max}$ and back to RT (RT$\rightarrow T_{a\ max} \rightarrow$ RT) in air, see Fig.2. This experiment clearly demonstrates that *the mechanical properties of the hard protective coating do not change during thermal cycling as far its structure remains unchanged.*

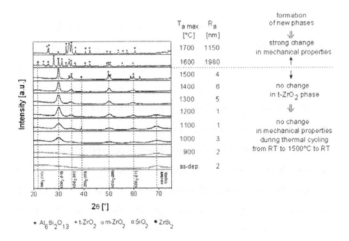

Fig.2. Evolution of X-ray diffraction structure and surface roughness R_a of 7000 nm thick $Si_{31}Zr_5O_{64}$ coating with $T_{a\,max}$ increasing from 900 to 1700°C. Adapted after reference [3].

2.1. Protection of substrate against oxidation by nanocomposite coatings

The enhanced protection of the substrate against oxidation by nanocomposite coatings is based on removing of a continuous connection between the coating surface and the substrate surface along grains boundaries, i.e. on the formation of amorphous coatings thermally stable up to high temperatures, see Fig.3. This schematic figure clearly shows that in the amorphous coating there is no contact of the external atmosphere with the substrate and so there is no reason for the substrate oxidation. The high temperature (high-T) protection of the substrate against oxidation by the amorphous protective coating fully depends on its thermal stability. The thermal stability is determined mainly by two factors - (1) the crystallization of amorphous material and (2) the diffusion of elements from (i) the substrate and (ii) oxidation atmosphere into the coating - which can stimulate the crystallization of the coating at $T < T_{cr}$. It means that new amorphous coatings thermally stable above 1000°C are necessary to be developed.

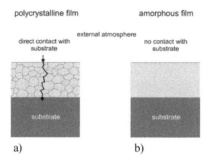

Fig.3. Principle of enhancement of the protection of substrate against oxidation by (a) crystalline nanocomposite coating and (b) amorphous coating [4]. The contact between the external atmosphere and the substrate is cancelled in amorphous coatings.

Recently, two new classes of *hard amorphous coatings* thermally stable above 1000 °C have been developed .

1. a-Si$_3$N$_4$/MeN$_x$ coatings with high content of a-Si$_3$N$_4$ phase; here Me=Ta, Zr, Ti, Mo, W, etc. [5-13]
2. a-Si-B-C-N coatings with strong covalent bonds [14-17]

These amorphous coatings were reactively sputtered using an unbalanced magnetron. They exhibit (i) high hardness ranging from ~20 to ~40 GPa and (ii) protect the substrate against oxidation in flowing air at high temperatures above 1000°C. Physical and mechanical properties of X-ray amorphous coatings strongly depend on (i) the thermal stability of individual phases from which they are composed and (ii) the diffusion of substrate elements into coating during thermal annealing.

The present state-of-the art in the field of the protection of the substrate against oxidation by hard non-oxide protective coatings based on nitrides is summarized in Fig.4. The oxidation resistance (OR) of the coating/substrate couple is characterized by the increase of the mass Δm after its annealing to a given temperature T_a. The value of T_a corresponding to the beginning of the sharp increase of Δm from $\Delta m \approx 0$ is the maximum temperature $T = T_{a\,max}$ at which the substrate is well protected against oxidation. The higher is the maximum temperature $T_{a\,max}$ the higher is the protection of substrate against oxidation.

Fig.4 clearly shows that the protection of substrate by the protective coating strongly depends on its structure. The crystalline coatings ensure a good oxidation protection only at low temperatures T (not exceeding ~1000°C) compared to the X-ray amorphous coatings. It is due to the fact that the crystalline coatings are composed of grains and always there is a direct contact between the external atmosphere at the coating surface and the substrate surface via grain boundaries, see Fig.3a. A small improvement in the OR can be achieved if an inter-granular glassy phase is formed. On the other hand, *the amorphous coatings protect the substrate against oxidation at much higher temperatures than the crystalline coatings and easily exceeds 1000°C by about 500 to 600°C.*

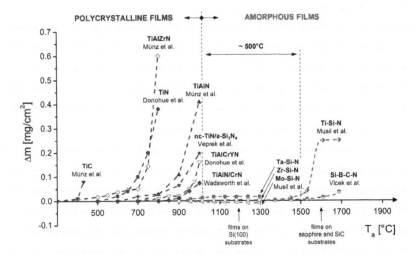

Fig.4. Oxidation resistance of selected hard binary, ternary, quaternary nitrides and amorphous Si_3N_4/MeN_x composite coatings characterized by the increase of the mass Δm of the coating/substrate couple with increasing annealing temperature T_a, i.e. $\Delta m = f(T_a)$ [1].

3.   HARD NANOCOMPOSITE COATINGS WITH ENHANCED TOUGHNESS

Up to recently, the attention was concentrated mainly on the hardness H of coatings, particularly on ways how (i) to enhance hardness H and (ii) to produce nanocomposites with H approaching or even exceeding that of the diamond. Extremely high hardness H of the film is not, however, needed for many applications. The hard films are brittle and their brittleness increases with increasing H. It means that the high brittleness of hard coatings strongly limits their practical utilization. It concerns mainly hard ceramic materials based on nitrides, oxides, carbides, borides, etc. and their combinations which are widely used in many applications either as protective or functional coatings. This is a main reason why now many labs over all world are trying to develop new ceramics with a low brittleness and simultaneously with sufficiently high (≥ 20 GPa) hardness. The coating material must exhibit not only the high hardness H but also a sufficient toughness because the film toughness can be for many applications more important than its hardness.

The way how to produce hard and tough coatings indicates the Hooke's law $\sigma = E.\varepsilon$; here, σ is the stress (load) and ε is the strain (deformation). If we need to form the material which exhibits a higher elastic deformation (a higher value ε) at a given value σ its Young's modulus E must be reduced. It means that coatings and materials with the lowest value of the Young's modulus E at a given hardness H (σ=const) need to be developed. Such coatings represent *a new generation of the advanced hard nanocomposite coatings*. Recent experiments indicate that the hardness H of hard, tough and/or resilient coatings ranges from about 15 to 25 GPa [18-22]. It is a sufficient hardness H for many applications. A reduction of the value of the Young's modulus E is a simple solution but a very difficult task. For more details see the reference [2].

The stress σ vs strain ε dependences for brittle, tough and resilient hard coatings are schematically displayed in Fig.5. Superhard materials are very brittle, exhibit almost no plastic deformation and very low strain $\varepsilon=\varepsilon_1$. Hard and tough materials exhibit both elastic and plastic deformation. The material withstanding a higher strain $\varepsilon_1 \ll \varepsilon \leq \varepsilon_{max}$ without its cracking exhibits a higher toughness. The hardness of tough materials is higher in the case when ε_{max} is achieved at higher values of σ_{max}. On the other hand, the fully resilient hard coatings exhibit compared with the superhard and tough materials a lower hardness H, no plastic deformation and 100% elastic recovery W_e (line 0A).

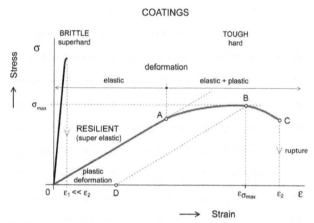

Fig.5. Schematic illustration of stress σ vs strain ε curves of superhard (brittle), hard (tough) and hard (resilient) coatings. Resilient coatings exhibit no plastic deformation (line 0A) [2].

4. HARD NANOCOMPOSITE COATINGS RESISTANT TO CRACKING

Recent experiments indicate that hard coatings resistant to cracking must simultaneously fulfill at least three conditions. They must have (1) the high ratio $H/E^* \geq 0.1$, (2) the high elastic recovery $W_e \geq 60\%$ and (3) the dense almost featureless microstructure [2]. The coatings fulfilling these requirements can be prepared by (i) the addition of a selected element in the base material and/or (i) the delivery of a sufficient energy E in the growing film. The effect of the element addition in the base material on its mechanical properties (H, E^*) and mechanical behaviour (H/E^*, W_e) is illustrated on properties of the Zr-Al-O films with (a) low ratio Zr/Al<1 and (b) high ratio Zr/Al>1, see Fig.6

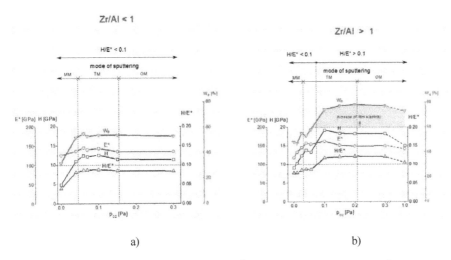

Fig.6. Hardness H, effective Young's modulus E^*, elastic recovery W_e and H/E^* ratio of the Zr-Al-O films with low ratio Zr/Al<1 and high ratio Zr/Al>1 sputtered on Si(100) substrate at $U_s=U_{fl}$, $T_s=500°C$, $p_T=1$ Pa, and $I_{da}=1$ and 2 A as a function of partial pressure of oxygen p_{O2}. MM, TM and OM are the metallic, transition and oxide modes of sputtering, respectively [21].

From Fig.6 it is seen that H, E^*, H/E^* and W_e of Zr-Al-O films can be controlled by the Zr/Al ratio and partial pressure of oxygen p_{O2} used in their sputtering. While the Zr-Al-O films with low ratio Zr/Al<1 exhibit only low ratio H/E^*<0.1, the Zr-Al-O films with high ratio Zr/Al >1 and high ratio $H/E^*≥0.1$ can be easily produced in wide range of p_{O2}. This experiment shows that the selection of the correct combination of the elemental composition of film and the deposition parameters used in its sputtering is of key importance in the case when the film with high ratio $H/E^*≥0.1$ is required to be produced.

Two Zr-Al-O coatings with low ratio H/E^*<0.1 (coating A) and high ratio $H/E^*≥0.1$ (coating B) were prepared to demonstrate the effect of the H/E^* ratio on the resistance of the Zr-Al-O film to cracking. The resistance of the coating to cracking was tested by bending. The coating is deposited on a metallic strip and the coated strip was bended along a fixed cylinder of diameter $\varnothing_{fc}$ using a moving cylinder of diameter $\varnothing_{mc}$ up to an angle α_c at which cracks in the coating occurred; more details are given in the reference [19]. Results of this experiment are given in Fig.7. The coating A with a low ratio H/E^*<0.1 easily cracks already after bending at the angle $\alpha=30°$; cracks are perpendicular to the direction of bending of the coated Mo strip. On the other hand, the coating B with a high ratio H/E^*>0.1 exhibits the strongly enhanced resistance to cracking and no cracks are created even after bending up to $\alpha=180°$; α is the angle of bending. The enhanced resistance to cracking of the coating B is due to the fact that the material with H/E^*>0.1 is tough and much more elastic ($W_e=75\%$) compared to the coating A with $H/E^*≤0.1$ whose material is brittle and less elastic ($W_e=52\%$). The load applied to the coating B is distributed over a wider area and thereby its resistance to cracking is enhanced.

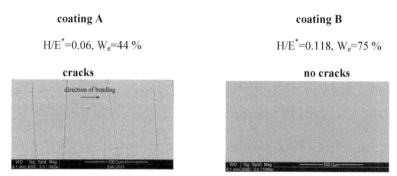

Fig.7. Surface morphology of ~3000 nm thick Zr-Al-O coating deposited on Mo strip after bending along steel cylinder of diameter $\varnothing_{fc} = 25$ mm [19].

Here, it is worthwhile to note that the Zr-Al-O films with Zr/Al>1 and H/E*>0.1 are nc-ZrO_2/a-Al_2O_3 composite composed of ZrO_2 nanograins embedded in the amorphous Al_2O_3 matrix. The cross-section microstructure of this film, which can be called as a glassy-like microstructure penetrated by irregular nanocolumns perpendicular to coating/substrate interface, is very dense. We believe that the microstructure is also responsible for the enhancement of the elasticity of the film and its resistance to cracking. This hypothesis is based on the fact that the films with columnar microstructure composed of densely packed columns very easily cracks, for instance see the reference [2,24]. It indicates that the high ratio H/E*>0.1 and the high elastic recovery $W_e \geq 60\%$ are necessary but not sufficient conditions ensuring the enhanced resistance of the film to cracking. The dense microstructure of the film is also of key importance for the enhancement of its resistance to cracking.

Recently, the resistance of the film against cracking was demonstrated in the following coating systems: (1) Zr-Al-O oxide/oxide nanocomposite coating [19,21], (2) Al-Cu-O oxide/oxide nanocomposite coating [18], (3) Al-O-N nitride/oxide nanocomposite coating [20] and (4) Si-Zr-O oxide/oxide nanocomposite coating [22]. All these coatings exhibit the high ratio H/E*≥ 0.1 and high elastic recovery $W_e \geq 60\%$. It indicates that the high ratio H/E*≥ 0.1 and high $W_e \geq 60\%$ are key conditions necessary for the formation hard films resistant to cracking. For more details see the papers [2,18-22].

In summary, it can concluded that a new task in the development of advanced hard nanocomposite coatings with enhanced toughness and enhanced resistance to cracking is to produce the coatings with (i) the low value of the Young's modulus E^* satisfying H/E*≥ 0.1 ratio, (ii) the high value of the elastic recovery W_e and (iii) the dense columnar free microstructure. The optimum microstructure of coating ensuring its enhanced resistance to cracking is not known so far and now is under intensive investigation. Hard, tough, elastic (HTE) nanocomposite coatings represent a new generation of advanced hard coatings. This new generation of HTE nanocomposite coatings is of huge potential for many applications, e.g. flexible electronics, flat panel displays, multifunctional coatings, advanced coatings for cutting tools, thermal barrier coatings, microelectromechanic systems (MEMS), protection of surfaces against cracking, etc.

5. TRENDS OF NEXT DEVELOPMENT

Main trends of the next development of new advanced hard thin films and coatings are the following: (i) the formation of new advanced nanocomposite coatings with enhanced hardness, toughness, elasticity and resistance to cracking in the oxide/oxide, oxide/nitride, nitride/nitride, nitride/boride, oxide/boride, nitride/carbide, boride/carbide, etc. nanocomposite systems and also in other material systems such as multi-element coatings, alloys, metallic glasses etc., (ii) the investigation of (a) the relationships between the resistance to cracking and the coating microstructure, (b) the thermal stability and (c) the oxidation, corrosion, wear resistance of new advanced nanocomposite coatings, (iii) the development of new advanced multi-function coatings based on nanocomposite coatings which are simultaneously hard, tough, elastic and resistant to cracking, (iv) the formation of high-temperature phases in hard coatings produced on unheated substrates and at low temperatures lower than 500°C, (v) the investigation of the electronic bonding between atoms in the nanocomposite and multilayered coatings, (vi) the formation of crystalline films on unheated thermally sensitive polymer substrates, e.g. for the flexible electronics, and (vii) the development of new sputtering sources and sputter deposition devices operating under new physical conditions. The principle of formation of crystalline films on unheated substrates is based on replacement of the conventional heating (T_s) with the atomic scale heating, see Fig.8. The energy necessary for the film crystallization is delivered to during its growth by bombarding and condensing particles. From Fig.8 it is also seen that new magnetrons operated at low ($\leq$0.1 Pa) pressures are needed to be developed.

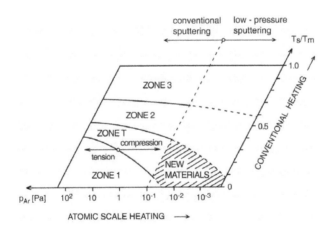

Fig.8. 2D Thornton's structural zone model extended into region of low-pressure sputtering. Here, T_s and T_m are the substrate temperature and the melting temperature of film material, respectively [23].

ACKNOWLEDGEMENTS

This work was supported in part by the Grant Agency of the Czech Republic GACR under project P108/12/0393.

REFERENCES

[1] J.Musil, P.Baroch and P.Zeman: Hard nanocomposite coatings. Present status and trends, Chapter 1 in the book "*Plasma Surface Engineering Research and its Practical Applications*", R.Wei (Ed.), Research Signpost Publisher, USA, 2008, 1-34.

[2] J.Musil: Hard nanocomposite coatings: Thermal stability, oxidation resistance and toughness, Surf.Coat.Technol. 207 (2012), 50-65.

[3] J.Musil, V.Šatava, P.Zeman and R.Čerstvý, Protective Zr-containing SiO_2 coatings resistant to thermal cycling in air up to 1400°C, Surf.Coat.Technol. 203 (2009), 1502-1507.

[4] J.Musil, P.Zeman: Hard amorphous a-Si_3N_4/MeN$_x$ nanocomposite coatings with high thermal stability and high oxidation resistance, Solid State Phenomena, Invited paper at the International Workshop on Designing of Interfacial Structures in Advanced Materials and Their Joints (DIS'06), May 18-20, 2006, Osaka, Japan and Solid State Phenomena (Designing of Interfacial Structures) 127 (2007), 31-36.

[5] H.Zeman, J.Musil and P.Zeman: Physical and mechanical properties of sputtered Ta-Si-N films with a high (≥40 at.%) content of Si, J.Vac.Sci.Technol. A 22(3) (2004), 646-649.

[6] J.Musil R.Daniel, P.Zeman and O.Takai: Structure and properties of magnetron sputtered Zr-Si-N films with a high (≥25 at.%) Si content, Thin Solid Films 478 (2005), 238-247.

[7] J.Musil, P.Dohnal and P.Zeman: Physical properties and high-temperature oxidation resistance of Si_3N_4/MoN$_x$ nanocomposite coatings with high (≥50 vol.%) of Si_3N_4 phase, J.Vac.Sci.Technol. B JUL-AUG (2005), 1568-1575.

[8] J.Musil, R.Daniel, J.Soldán and P.Zeman: Properties of reactively sputtered W-Si-N films, Surf.Coat.Technol. 200 (12-13)(2006), 3886-3895.

[9] P.Zeman and J.Musil: Difference in high-T oxidation resistance of amorphous Zr-Si-N and W-Si-N films with a high Si content, Applied Surface Science 252 (2006), 8319-8325.

[10] P.Zeman, J.Musil and R.Daniel: High-temperature oxidation of Ta-Si-N films with a high Si content, Surf.Coat.Technol. 200 (2006), 4091-4096.

[11] R.Daniel, J.Musil, P.Zeman and C.Mitterer: Thermal stability of magnetron sputtered Zr-Si-N films, Surf.Coat.Technol. 201 (2006), 3368-3376.

[12] J.Musil, P.Zeman, P.Dohnal and R.Čerstvý: Ti-Si-N nanocomposite films with a high content of Si, Plasma Processes and Polymers Vol.4, Issue S1 (2007), S574-S578.

[13] J.Musil, J.Vlček and P.Zeman: Advanced amorphous non-oxide coatings with oxidation resistance above 1000°C, Special Issue on NANOCERAMICS, Advances in Applied Ceramics 107(3) (2008), 148-154.

[14] J.Vlček, Š.Potocký, J.Čížek, J.Houška, M.Kormunda, P.Zeman, V.Peřina, J.Zemek, Y.Setsuhara and S.Konuma: Reactive magnetron sputtering of hard Si-B-C-N films with a high-temperature oxidation resistance, J. Vac. Sci. Technol. A 23(6) (2005), 1513-1521.

[15] J.Houška, J.Vlček, S.Hřeben, M.M.M.Bilek and D.R.McKenzie: Effect of B and Si/C ratio on high-temperature stability of Si-B-C-N materials, Europhys. Lett. 76 (2006), 512-518.

[16] J.Houška,J.Vlček, Š.Potocký and V.Peřina: Influence of substrate bias voltage on structure and properties of hard Si-B-C-N films prepared by reactive magnetron sputtering Diamond Relat. Matter. 16, (2007), 29-36.

[17] J.Vlček, S.Hřeben, J.Kalaš, J.Čapek, P.Zeman, R.Čerstvý, V.Peřina and Y.Setsuhara: Magnetron sputtered Si-B-C-N films with high oxidation resistance and thermal stability in air at temperatures above 1500°C, J.Vac.Sci.Technol. A 26(5) (2008), 1101-1108.

[18] J.Blažek, J.Musil, P.Stupka, R. Čerstvý and J.Houška: Properties of nanocrystalline Al-Cu-O films reactively sputtered by dc pulse dual magnetron, Applied Surface Science 258 (2011), 1762-1767.

[19] J.Musil, J.Sklenka and R.Čerstvý: Transparent Zr-Al-O oxide coatings with enhanced resistance to cracking, Surf.Coat.Technol. 206 (2011), 2105-2109.

[20] J.Musil, M.Meissner, R.Jilek, T.Tolg, R.Čerstvý: Two-phase single layer Al-O-N nanocomposite films with enhanced resistance to cracking, Surf.Coat.Technol. 206 (2012), 4230-4234.

[21] J.Musil, J. Sklenka, R.Čerstvý, T.Suzuki, M.Takahashi, T.Mori: The effect of addition of Al in ZrO_2 thin film on its resistance to cracking, Surf.Coat.Technol. 207 (2012), 355-360.

[22] Si-Zr-O nanocomposite coatings, unpublished results

[23] J. Musil: Low-pressure magnetron sputtering, Vacuum 50(3-4) (1998), 363-372.

[24] Y.T.Pei, D.Galvan, J.Th.M.De Hosson: Nanostructure and properties of TiC/a-C:H composite coatings, Acta Materialia 53 (2005), 4505-4521.

PREPARATION OF EPITAXIALLY GROWN Cr-Si-N THIN FILMS BY PULSED LASER DEPOSITION

T. Endo, K. Suzuki, A. Sato, T. Suzuki, T. Nakayama, H. Suematsu and K. Niihara

Extreme Energy-Density Research Institute, Nagaoka University of Technology, Nagaoka, Japan

ABSTRACT

Cr-Si-N thin films having nanosized CrN crystallites with B1 (NaCl type) structure have higher hardness and tribological properties than those of CrN. It is thought that solubility of Si atoms in the CrN lattice with six coordination is limited, since Si is easy to form four coordination. However, whether Si is contained in the nanosized CrN crystal has not been clear. In this work, epitaxially grown Cr-Si-N thin films with B1 structure have deposited on MgO(100) substrates using pulsed laser deposition. The laser was focused on a rotating target that a Si wafer was attached on a Cr disk. The ratio of surface area of Si for Cr (S_R) is controlled. Nitrogen plasma from an RF radical source was introduced in the deposition chamber. The prepared thin films were characterized by X-ray diffraction, energy dispersive X-ray spectroscopy and electron energy loss spectroscopy. In these results, both samples of $S_R = 0$ and 10%, $i.e.$ CrN and Cr-Si-N, have B1 structure and epitaxial grown on MgO substrate. The Si content in these thin films were 0 and 5.0 at.%. Hardness of CrN and Cr-Si-N measured by nanoindentation were 29 and 36 GPa, respectively.

INTRODUCTION

Hard thin film coatings have been applied to wide fields such as cutting tools, mechanical parts and molds for protecting these base materials against the damages from heat, oxidation and abrasion[1]. Titanium nitride (TiN), chromium nitride (CrN) and diamond-like carbon (DLC) are used for the hard coatings. Although DLC is excellent in hardness, DLC has the problems in structural change at high temperature[2] and high reactivity with iron. For these reasons, TiN and CrN are used for cutting tools. CrN has good oxidation- and wear-resistance but its hardness is less than that of TiN, since CrN characteristics have been improved by adding other elements such as $(Cr,Al)N$[3], $Cr(N,O)$[4], Cr-Al-N-O[5], Cr-Si-N[6] and Cr-Si-N-O[7].

Thin films of Cr-Si-N typically consist of two phases; nanosized CrN crystallites with B1 (NaCl type) structure and an amorphous SiN_x phase, thus Cr-Si-N can also be written in nc-CrN/a-SiN_x[6,8]. The hardness of Cr-Si-N increases to maximum hardness depending on Si content up to a few percent and it decreases at more than above Si content gradually[6,8]. This tendency is also known in Cr-Si-N-O thin films[6]. The reason why the two phases nc-CrN/a-SiN$_x$ are formed is that Si atoms is easy to form four coordination due to its sp^3 hybridized orbital while the CrN lattice is six coordination[9]. However, whether Si is contained in the nanosized CrN crystal has not been cleared.

If Cr-Si-N crystals which have dissolved Si atoms into the B1 lattice can be prepared, the improvement of hardness can be expected with the distortion of the crystal lattice such as Cr(N,O)[3]. It has been reported that CrN is epitaxially grown on MgO (001) substrates[10]. Pulsed laser deposition (PLD) is a good method to control the composition of thin films, which is ideal to deposit Cr-Si-N with precise Si content control. However, there are no reports that Cr-Si-N was prepared by PLD. In addition, the solubility of Si atoms into CrN lattice has not been investigated. In this work, we prepared Cr-Si-N thin films on Si(100) and MgO(100) substrates by PLD and investigated the solid solution of Si atoms into CrN lattice for improvement of the hardness.

EXPERIMENTAL

Figure 1 shows a schematic drawing of the pulsed laser deposition (PLD) apparatus used for preparing the thin films. The ablation plasma was produced by irradiating third harmonic of an Nd: yttrium aluminum garnet laser (355 nm) onto a rotating target, in which a single crystal Si(100) wafer was attached on a Cr (99.9% purity) disk. The ratio of surface area of Si for Cr (S_R) was 0 and 10%. The laser was electro-optically Q-switched by a Pockels cell to produce intense pulses of short duration (7 ns) and a laser-pulse repetition rate was 10 Hz. The area irradiated by the pulsed laser onto the target was 0.79 mm^2 and the energy density was set as 1.7 J/cm^2. The deposition time was 24 hours. The thicknesses of prepared thin films were approximately 255-275 nm. The substrates were single crystals of Si(100) and MgO(100) cleaned by steeping in hydrofluoric acid for about 10 minutes. Both substrates were used in a single experiment and thin films were deposited on the substrates simultaneously. The deposition surface area was about 100 mm^2. The substrates placed at a distance of 50 mm from the target. The substrates temperature was controlled at 973 K using an infrared lamp heater.

The chamber was first evacuated to a pressure of under 1.0×10^{-6} Pa using a rotary pump and a turbo molecular pump, and the chamber was then filled with nitrogen plasma (>99.99995 vol.% purity) from an RF radical source (model RF-4.5, SVT Associates, USA). Both continuous pumping and introduction of N plasma were carried out during the deposition. The thin films were prepared under a fixed total pressure of 1.5×10^{-2} Pa.

The crystal structures of the thin films were studied by X-ray diffraction (XRD) using

CuKα radiation (0.154 nm). θ/2θ method was applied to the films on MgO(100) and Si(100) substrate, and grazing incidence (incidence angle α = 1.5°) and phi scan were applied to only the films on MgO(100) substrate. The chosen reflection for the phi scan of the films was (111), whose diffraction angle was determined by the lattice constant obtained from the θ/2θ scan data. For comparison, the MgO(100) substrate was characterized with this configuration and no peaks were detected. This preliminary measurement ensured that diffraction only from thin films was detected. The compositions of the thin films were determined by energy dispersive X-ray spectrometry (EDS) and electron energy loss spectroscopy (EELS). The hardness of the thin films was measured by nanoindentation testing (Nanoindenter G200, Agilent Technology, USA) under a load of 0.6 mN using a Berkovich indenter. The microstructures of the thin films were observed using a field emission transmission electron microscope (FE-TEM) with a 200 kV acceleration voltage. The TEM samples were made by a focused ion beam (FIB) apparatus.

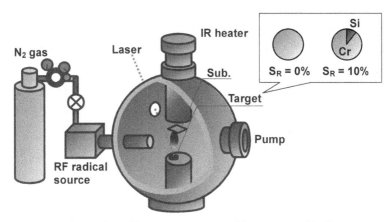

Figure 1. Schematic drawing of the PLD apparatus used for preparing thin films.

RESULTS AND DISCUSSION

Phase identification

Figure 2(a) shows XRD patterns of $S_R = 0\%$ (CrN) thin films. In the $\theta/2\theta$ scan data of the thin film on MgO(100) substrate, a peak due to CrN(200) was shifted to a low angle and closer to the MgO(100) peak. Peaks other than CrN(200) were not detected. When diffraction from the thin film on MgO(100) was measured at 42-45 degree, a Cu plate absorber in front of X-ray detector was used. From results of grazing incidence, no peaks were observed, which suggests high crystal orientation of this film on MgO(100) substrate. Figure 2(b) shows the result of phi scan, and only four peaks were observed at every 90°. These peaks derive from a four-fold symmetry (111) with B1 structure. Diffraction angle of (111) was calculated by (200) angle in $\theta/2\theta$. From these results, it was thought that the CrN film on MgO(100) substrate was epitaxial grown. This thin film was polycrystalline in contrast to the single crystal film on MgO(100) substrate.

Figure 3(a) and 3(b) show XRD patterns of $S_R = 10\%$ (Cr-Si-N) films. Both patterns were similar to those of $S_R = 0\%$ film. From these results, it was suggested that the epitaxially grown Cr-Si-N thin films were deposited on the MgO(100) substrate.

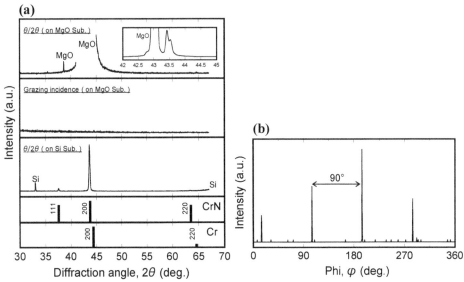

Figure 2. (a)XRD pattern of $S_R = 0\%$ (CrN) thin films; and (b)Phi scan of thin film.

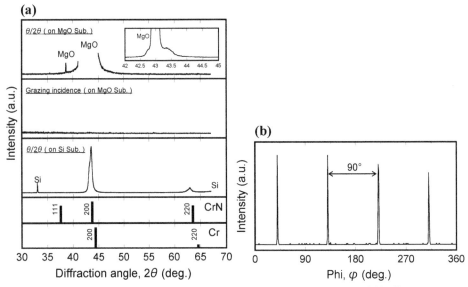

Figure 3. (a)XRD pattern of $S_R = 10\%$ (Cr-Si-N) thin films; and (b)Phi scan of thin film.

Microstructure

Figure 4(a) shows the high resolution TEM (HRTEM) images and selected area diffraction (SAD) patterns of the $S_R = 0\%$ thin film on MgO substrate. From the SAD pattern, the electron beam was incidented parallel to the [001] direction of the thin film. This pattern was unchanged in the 0.2 × 5 μm area. From the HRTEM image, the {200} planes of the CrN lattice were visible, which were parallel to the {200} planes of the MgO substrate. It was found that the CrN thin film was epitaxially grown on the MgO(100) substrate and could be of a single crystal. Figure 4(b) shows the same images of the thin film deposited on Si substrate. The grain boundaries were clearly observed in HRTEM image, and SAD showed that the thin film was polycrystalline.

Figure 5(a) and 5(b) show the images of $S_R = 10\%$ thin film on MgO and Si substrate, respectively. These results were similar those of the $S_R = 0\%$ film. This thin film on MgO substrate is likely to be a Cr-Si-N single crystal.

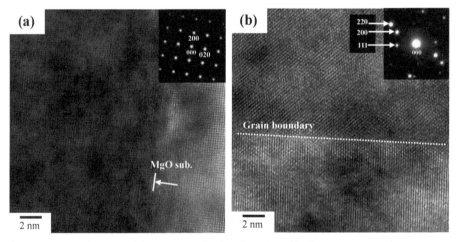

Figure 4. HRTEM images and SAD patterns for $S_R = 0\%$ (CrN) thin film on (a)MgO substrate and (b) Si substrate.

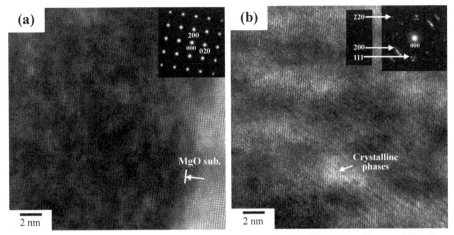

Figure 5. HRTEM images and SAD patterns for S_R = 10% (Cr-Si-N) thin film on (a)MgO substrate and (b) Si substrate.

Compositions and Hardness

Table 1 shows the film compositions measured by EELS and EDS. EELS was applied to the comparison of Cr and N contents, and EDS were applied to that of Cr and Si contents. The all composition (Cr, Si and N) in the films was calculated from the results of EELS and EDS.

Table 1. Compositions of thin films measured by EELS and EDS

S_R (%)		MgO Sub.			Si Sub.		
		Cr	Si (at. %)	N	Cr	Si (at. %)	N
0	EELS	45	-	55	51	-	49
	All composition	45	-	55	51	-	49
10	EELS	44	-	56	49	-	51
	EDS	89	11	-	88	12	-
	All composition	42	5	53	46	6	48

Although the Cr content was slightly less than that of N in $S_R = 0\%$ thin film on MgO substrate, the thin film epitaxially grown on MgO substrate may have cation vacancies. In the Cr-Si-N for $S_R = 10\%$, the thin film on MgO substrate shows similar nonstoichiometry. The EDS data shows presence of Si in the thin film on MgO substrate. From Fig.5(a), no grain boundary was seen so that Si must be dissolved in the crystal to form (Cr,Si)N phase.

Figure 6 shows the hardness of thin films measured by a nanoindenter. The film on the MgO substrate showed a higher hardness than that of the film on Si substrate in $S_R = 0$ and 10%. In the polycrystalline thin film on Si substrate, conventional result that hardness increases with increasing Si content[7,8] was confirmed. The hardness was increased from 23 to 29 GPa. Furthermore, in the thin film on MgO substrate, hardness increases with increasing Si content. The hardness was increased from 29 to 36 GPa. This result could be supposed the effect of the obstruction of dislocation movement (solid solution hardening) for solubility of Si atoms in the CrN lattice. As the result, Cr-Si-N thin film on MgO substrate ($S_R = 10\%$, Si content = 5.0 at.%) showed the highest hardness 36 GPa.

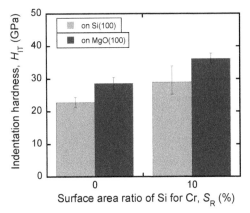

Figure 6. Indentation hardness for the thin films of $S_R = 0$ and 10% on MgO and Si substrate.

Conclusion

From the results of deposition of Cr-Si-N thin films by PLD on the characterizations, it was found that CrN and Cr-Si-N was formed as B1 crystals epitaxially grown on MgO(100) substrates and these thin films had Si content of 0 and 5.0 at.%, respectively. On the other hand, CrN and Cr-Si-N thin films on Si(100) substrates were polycrystalline. The Cr-Si-N thin film on MgO(100) substrate showed that the Cr content was less than that of N. This result suggested that (Cr,Si)N phase was formed. Furthermore, adding Si into CrN on MgO(100) substrate, the hardness was increased from 29 to 36 GPa. This result implied the effect of solid solution hardening for solubility of Si atoms in the CrN lattice as (Cr,Si)N.

ACKNOWLEDGEMENT

This work was supported by Grant-in-Aid for Scientific Research 22686069.

REFERENCES

[1] B. Navinsek, P. Panjan and I. Milosev: Industrial Applications of CrN (PVD) Coatings, Deposited at High and Low Temperatures, *Surface and Coatings Technology*, **97**, 182-191 (1997)

[2] Won Jae Yang, Tohru Sekino, Kwang Bo Shim, Koichi Niihara and Keun Ho Auh: Microstructure and Tribological Properties of SiO_x/DLC Films Grown by PECVD, *Surface & Coatings Technology*, **194**, 128-135 (2005).

[3] In-Wook Park, Dong Shik Kang, John J. Moore, Sik Chol Kwon, Jong Joo Rha and Kwang Ho Kim: Microstructures, Mechanical Properties, and Tribological Behaviors of Cr-Al-N, Cr-Si-N,

and Cr-Al-Si-N Coatings by a Hybrid Coating System, *Surface & Coatings Technology*, **201**, 5223-5227 (2007).

[4] J. Inoue, H. Saito, M. Hirai, T. Suzuki, H. Suemastsu, W. Jiang and K. yatsui: Mechanical Properties and Oxidation Behavior of Cr-N-O Thin Films Prepared by Pulsed Laser Deposition, *Transactions of the Materials Research Society of Japan*, **28 (2)**, 421-424 (2003).

[5] Makoto Hirai, Hajime Saito, Tsuneo Suzuki, Hisayuki Suematsu, Weihua Jiang and Kiyoshi Yatsui: Oxidation Behavior of Cr-Al-N-O Thin Films Prepared by Pulsed Laser Deposition, *Thin Solid Films*, **407**, 122-125 (2002).

[6] E. Martinez, R. Sanjines, A. Karimi, J. Esteve and F. Levy: Mechanical Properties of Nanocomposite and Multilayered Cr-Si-N Sputtered Thin Films, *Surface & Coatings Technology*, **180-181**, 570-574 (2004).

[7] Jun Shirahata, Aoi Sato, Kazuma Suzuki, Tetsutaro Ohori, Hiroki Asami, Tsuneo Suzuki, Tadachika Nakayama, Hisayuki Suematsu and Koichi Niihara: Mechanical Properties and Microstructures of Silicon Doped Chromium Oxynitride Thin Films, *Journal of the Japan Institute of Metals*, **75**, 97-103 (2011)

[8] Jong Hyun Park, Won Sub Chung, Young-Rae Cho and Kwang Ho Kim: Synthesis and Mechanical Properties of Cr–Si–N Coatings Deposited by a Hybrid System of Arc Ion Plating and Sputtering Techniques, *Surface & Coatings Technology*, **188-189**, 425-430 (2004).

[9] Walter A. Harrison, Electronic structure and the properties of solids, *Dover Publications*, New York, 1989.

[10] Kei Inumaru, Kunihiko Koyama, Naoya Imo-oka and Shoji Yamanaka: Controlling the Structural Transition at the Néel Point of CrN Epitaxial Thin Films Using Epitaxial Growth, *PHYSICAL REVIEW B*, **75**, 054416 (2007).

INFLUENCE OF OXYGEN CONTENT ON THE HARDNESS AND ELECTRICAL
RESISTIVITY OF Cr(N,O) THIN FILMS

Aoi Sato, Toshiyuki Endo, Kazuma Suzuki, Tsuneo Suzuki, Tadachika Nakayama,

Hisayuki Suematsu and Koichi Niihara,

Extreme Energy-Density Research Institute, Nagaoka University of Technology

1603-1 Kamitomioka, Nagaoka, Niigata 940-2188, JAPAN

ABSTRACT

     Chromium oxynitride thin films were deposited on Si (100) substrates by pulsed laser deposition. In order to change the oxygen content of thin films, oxygen partial pressure (Po_2) was varied from 5.0×10^{-5} Pa to 10×10^{-5} Pa. Compositions of the thin films determined by Rutherford backscattering spectroscopy and electron energy loss spectroscopy. The oxygen content (C_O) of the thin films was increased from 0 to 40 mol% with increasing Po_2. Phase identification of the thin films analyzed by Fourier-transform infrared spectroscopy and X-ray diffraction. The thin films with only the NaCl-type Cr(N,O) phase were prepared at $Po_2 = 7.5 \times 10^{-5}$ Pa or less. The hardness of the Cr(N,O) thin films measured by nano-indenter was increased from 19 to 32 GPa with increasing C_O. Electrical resistivity of Cr(N,O) thin films measured by a four probe method at room temperature was increased from 10^{-4} to 10^1 Ωcm with increasing C_O.

INTRODUCTION

     Hard thin films are used in many applications such as auto-parts, mold tools and golf clubs. The reason is to improve those durability and performance. Above all, hard thin film has a key role for cutting tools. The cutting technology made rapid progress by development of titanium aluminum nitride (Ti-Al-N) thin film superior in hardness and oxidation resistance [1] and diamond-like carbon (DLC) thin film superior in hardness and lubricity [2, 3]. However recently, the cutting technology has been shifting from wet processes using cutting oil to dry processes in light of ecology. Therefore, characteristic improvement such as hardness, oxidation resistance and wear resistance in hard thin films is demanded.

The chromium nitride (CrN) is a superior coating material with high oxidation resistance, a low friction coefficient and good protection against welding. However, its application was limited, because the CrN thin film typically has lower hardness than those of other hard materials. We focused on this point, and have originally developed chromium oxynitride (Cr(N,O)) thin films [5-7] and chromium magnesium oxynitride ((Cr,Mg)(N,O)) thin films [8, 9] by pulsed laser deposition (PLD), and confirmed that those hardness and oxidation resistance were improved. However, since those thin films were produced using residual oxygen and leak gas in a chamber, in the previous studies, reproducibility was limited.

In this study, we prepared Cr(N,O) thin films with precise controlling of oxygen content in thin films. In addition, electrical characteristics of the thin films were revealed. Although, electrical conductivity of CrN thin films, were already reported by Gall et al. [10] , Constantin et al. [11] and Inumaru et al. [12] , that of Cr(N,O) thin films have not been reported. Therefore, the purpose of this study is to investigate the influence of oxygen content on hardness and electrical resistivity of Cr(N,O) thin films.

EXPERIMENTAL

Chromium oxynitride thin films were deposited on single crystal Si (100) substrates by PLD method. Figure 1 showed the experimental setup. The ablation plasma was produced by irradiating the Nd: yttrium aluminum garnet laser beam (355 nm) onto a Cr (purity: 99.9 %) target. The laser was electro-optically Q-switched by a Pockels cell to produce intense, short pulses ($\sim$ 7 ns). The laser beam was focused by a lens, and the energy density was adjusted to 1.7 J/cm^2 on the target surface. The deposition was performed for 5 hours using a pulse repetition rate of 10 Hz. The substrate temperature was controlled at 973 K by infrared ray lamp heater. The substrate was placed at a distance of 45 mm from the target. Before the deposition, the chamber was evacuated to less than 2.5×10^{-5} Pa using a rotary pump and a turbo molecular pump. Then oxygen gas (purity: 99.99995 %) was introduced in the chamber. The reading of a vacuum gauge was defined as O_2 partial pressure (Po_2). In order to change the oxygen content of thin films, Po_2 was varied from 5.0×10^{-5} Pa to 10×10^{-5} Pa. After O_2 addition, nitrogen plasma (purity: 99.99995 %) from a radio-frequency (RF) radical source was supplied. Samples were prepared at a total pressure of 1.5×10^{-2} Pa in a mixed atmosphere of O_2 and N plasma. Table 1 showed typical experimental conditions.

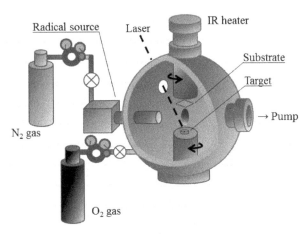

Figure 1. Experimental setup.

Table 1. Typical experimental conditions.

Target	Cr (purity: 99.9 %)
Substrate	Si (100)
Sub. temperature, T_{sub} / K	973
Distance (target-sub.), d_{TS} / mm	45
Base pressure, P_B / Pa	$< 2.5 \times 10^{-5}$
O_2 partial pressure, Po_2 / Pa	$5.0 - 10 \times 10^{-5}$
Atmosphere	O_2 + N plasma
Working pressure, P_W / Pa	1.5×10^{-2}
Energy density / Jcm^{-2}	1.7
Deposition time / hour	5

The thin film compositions were measured by Rutherford backscattering spectroscopy (RBS) and electron energy loss spectroscopy (EELS). Since measurement accuracy of light elements by RBS was limited, the cation to anions composition was measured. By EELS, N and O compositions were accurately determined. The chemical bonding states were examined by Fourier-transform infrared (FT-IR) spectroscopy. The crystal structures of the thin films were investigated by X-ray diffraction (XRD) using Cu Kα radiation (0.15418 nm) under operating conditions of 50 kV and 300 mA. The oxygen content was defined as C_O. The indentation hardness (H_{IT}) of the thin films was measured by nano-indenter (Fischer Instruments, HM2000) with a load of 0.07 mN using a Berkovich indenter. The measurement system of electrical resistance (R) was constructed with Lab VIEW, and it was measured by four probe method at room temperature. The electrical resistivity (ρ) was calculated with the following expression (1),

$$\rho = R \times \frac{w}{l} \times t .$$
(1)

Here, w is the sample width, l is the distance between probes for voltage measurement and t is the film thickness. The film thickness measured in a scanning electron microscope (SEM) was approximately 100 nm.

RESULTS AND DISCUSSION

Figure 2 shows compositions of the thin films determined by RBS and EELS. The oxygen content of the thin films was increased with increasing Po_2. This demonstrates that the oxygen content of the thin films can be controlled by varying Po_2. In addition, not only nitrogen content but also chromium content decreased with increasing Po_2. Therefore, vacancy would be included into chromium sites in the Cr(N,O) thin film.

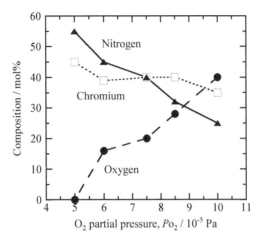

Figure 2. Compositions of the thin films determined by RBS and EELS.

Figure 3 shows FT-IR spectra of the thin films with various Po_2. The bottom of Fig. 3 is reference data for CrN and Cr_2O_3 [14] . Absorption peaks attributed to Cr-N bond were confirmed in the thin films prepared at $Po_2 = 7.5 \times 10^{-5}$ Pa or less. In contrast, absorption peaks attributed to not only Cr-N bond but also Cr-O bond were confirmed in the thin films of $Po_2 = 8.5 \times 10^{-5}$ Pa or more. Figure 4 shows XRD patterns of the thin films with various Po_2. The bottom of Fig. 4 is powder diffraction data for CrN and Cr_2O_3 from the International Center for Diffraction Data. Diffraction peaks attributed to CrN of the B1 (NaCl-type) structure were confirmed in samples prepared at $Po_2 = 8.5 \times 10^{-5}$ Pa or less. In contrast, diffraction peaks attributed to not only CrN but also Cr_2O_3 were observed in the sample prepared at $Po_2 = 10 \times 10^{-5}$ Pa. From the results in Figs. 3 and 4, tiny amount of crystalline and/or amorphous Cr_2O_3 were included in the sample of $Po_2 = 8.5 \times 10^{-5}$ Pa. Since samples prepared at $Po_2 = 7.5 \times 10^{-5}$ Pa or less keep B1 structure regardless of increasing C_O, the oxygen atoms substituted for nitrogen atoms in the structure. Moreover, samples prepared at $Po_2 = 8.5 \times 10^{-5}$ Pa or more consists of two phases of Cr(N,O) and Cr_2O_3. Thus, Cr(N,O) thin films were prepared by PLD method, and the limit of Po_2 was thought to be $Po_2 = 7.5 - 8.5 \times 10^{-5}$ Pa. Figure 5 shows the lattice constant of Cr(N,O) thin films. Those values were calculated with Bragg's law from the (111) and (200) diffraction positions in

Fig. 4. The lattice constant was linearly decreased with increasing Po_2. Therefore, a mechanical characteristic improvement is expected.

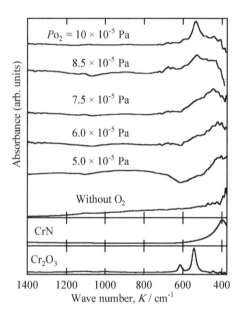

Figure 3. FT-IR spectra of the thin films with various Po_2.

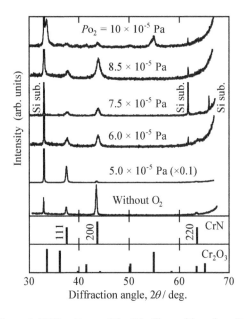

Figure 4. XRD patterns of the thin films with various Po_2.

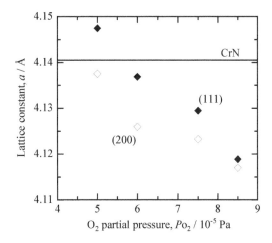

Figure 5. Lattice constant of Cr(N,O) thin films.

Figure 6 shows the indentation hardness of the thin films as a function of C_O. The top of Fig. 6 is the summary of phase identification by XRD and FT-IR. The hardness of Cr(N,O) thin films was increased from 19 to 32 GPa with increasing C_O. It was suggested that the cause was solution hardening by substituting oxygen atoms with nitrogen atoms. In contrast, the hardness of samples containing two phases decreased, because Cr_2O_3 is softer material than CrN.

Figure 7 shows electrical resistivity of the thin films as a function of C_O. As same as Fig. 6, the results of phase identification by XRD and FT-IR are shown on the top of Fig. 7. It was revealed that electrical resistivity of the Cr(N,O) thin films was increased from 10^{-4} to 10^{1} Ωcm with increasing C_O. Increasing electrical resistivity of the Cr(N,O) thin films is likely to be caused by strong electron correlation, because 3d transition metal monoxides except for TiO and VO indicated insulator property, and Cr_2O_3 is a Mott insulator. In order to reveal the cause, it is necessary for additional analyses such as carrier electron and density of state.

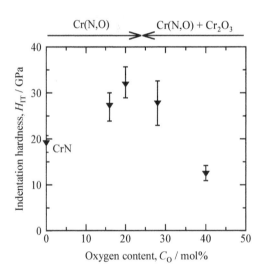

Figure 6. Indentation hardness of the thin films as a function of C_O.

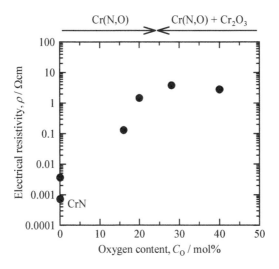

Figure 7. Electrical resistivity of the thin films as a function of C_O.

CONCLUSION

The oxygen content of the thin films was controlled by varying Po_2. From the results of phase identification by FT-IR and XRD, Cr(N,O) thin films were prepared by a PLD method, and the limit of Po_2 was thought to be $Po_2 = 7.5 - 8.5 \times 10^{-5}$ Pa. The hardness of Cr(N,O) thin films was increased from 19 to 32 GPa with increasing C_O. It was suggested that the cause was solution effect by substituting oxygen atoms with nitrogen atoms. Electrical resistivity of the Cr(N,O) thin films was increased from 10^{-4} to 10^{1} Ωcm with increasing C_O. The increasing electrical resistivity of the Cr(N,O) thin films is likely to be caused by strong electron correlation. Thus, the influence of oxygen content on hardness and electrical resistivity of Cr(N,O) thin films was revealed.

ACKNOWLEDGEMENT

This work was supported by Grant-in-Aid for Scientific Research 22686069.

REFERENCES

[1] T. Ikeda and H. Satoh: "High-Temperature Oxidation and Wear Resistance of Ti-Al-N Hard Coatings Formed by PVD Method", *J. Japan Inst. Metals*, **57**, 919-925 (1993)

[2] J. Robertson: "Properties of Diamond-like Carbon", *Surf. Coat. Technol.*, **50**, 185-203 (1992)

[3] T. Hasebe, Y. Matsuola, H. Kodama, T. Saito, S. Yohena, A. Kamijo, N. Shiraga, M. Higuchi, S. Kuribayashi, K. Takahashi, T. Suzuki: "Lubrication Performance of Diamond-like Carbon and Fluorinated Diamond-like Carbon Coatings for Intravascular Guide Wires", *Diamond Relat. Mater.*, **15**, 129-132 (2006)

[4] J. Vetter, E. Lugscheider and S. S. Guerreiro: "(Cr,Al)N Coatings Deposited by the Cathodic Vacuum Arc Evaporation", *Surf. Coat. Technol.*, **98**, 1223-1239 (1998)

[5] T. Suzuki, H. Saito, M. Hirai, H. Suematsu, W. Jiang and K. Yatsui: "Preparation of Cr(N,O) Thin Films by Pulsed Laser Deposition", *Thin Solid Films*, **407**, 118-121 (2002)

[6] J. Inoue, H. Saito, M. Hirai, T. Suzuki, H. Suematsu, W. Jiang and K. Yatsui: "Mechanical Properties and Oxidation Behavior of Cr-N-O Thin Films Prepared by Pulsed Laser Deposition", *Trans. Mater. Res. Soc. Jpn.*, **28**, 421-424 (2003)

[7] T. Suzuki, J. Inoue, H. Saito, M. Hirai, H. Suematsu, W. Jiang and K. Yatsui: "Influence of Oxygen Content on Structure and Hardness of Cr-N-O Thin Films Prepared by Pulsed Laser Deposition", *Thin Solid Films*, **515**, 2161-2166 (2006)

[8] H. Asami, J. Inoue, M. Hirai, T. Suzuki, T. Nakayama, H. Suematsu, W. Jiang, and K. Niihara: "Hardness Optimization of (Cr,Mg)(N,O) Thin Films Prepared by Pulsed Laser Deposition", *Advanced Materials Research*, **11**, 311-314 (2006)

[9] H. Asami, J, Inoue, T. Suzuki, T. Nakayama, H. Suematsu, W. Jiang and K. Niihara: "Oxidation Behavior of Cr-Mg-N-O Thin Films Prepared by Pulsed Laser Deposition", *J.Jpn. Soc. Powder Powder Metallurgy*, **54**, 198-201 (2006)

[10] D. Gall, C.-S. Shin, R. T. Haasch, I. Petrov and J. E. Greene: "Band Gap in Epitaxial NaCl-structure CrN(001) Layers", *J. Appl. Phys.*, **91**, 5882-5886 (2002)

[11] C. Constantin, M. B. Haider, D. Ingram and A. R. Smith: "Metal/semiconductor Phase Transition in Chromium Nitride(001) Grown by Rf-plasma-assisted Molecular-beam Epitaxy", *Appl. Phys. Lett.*, **85**, 6371-6373 (2004)

[12] K. Inumaru, K. Koyama, N. Imo-oka and S. Yamanaka: "Controlling the Structural Transition at the Néel Point of CrN Epitaxial Thin Films Using Epitaxial Growth", *PHYSICAL REVIEW B*, **75**, 054416 (2007)

[13] J. Shirahata, T. Ohori, H. Asami, T. Suzuki, T. Nakayama, H. Suematsu and K. Niihara: "Fourier-Transform Infrared Absorption Spectroscopy of Chromium Nitride Thin Film", *Jpn. J. Appl. Phys.*, **50**, 01BE03 (2011)

NANOCOMPOSITE MO-CU-N COATINGS DEPOSITED BY REACTIVE MAGNETRON SPUTTERING PROCESS WITH A SINGLE ALLOYING TARGET

Duck Hyeong Jung[12], Caroline Sunyong Lee[2], Kyoung Il Moon[1]
[1]Incheon Division, Korea Institute of Industrial Technology, Incheon 406-840, Korea
[2]Hanyang University, Ansan

ABSTRACT

In this study, it has been tried to make the single Mo-Cu alloying targets with the Cu content showing the best surface hardness that was determined by investigation on the coatings with the Mo and Cu double target process. The single alloying targets were prepared by powder metallurgy methods such as mechanical alloying and spark plasma sintering. Subsequently, the nanocomposite coatings were prepared by reactive magnetron sputtering process with the single alloying targets in Ar+N$_2$ atmosphere. The microstructure changes of the coatings were investigated by using XRD, SEM and EDS and TEM. The mechanical properties of Mo-Cu-N coatings were evaluated by using nanoindentor. In this study, the nano-composite MoN-Cu coatings prepared using an alloying target was eventually compared with the coatings from the multiple targets.

INTRODUCTION

Recently, industry demands a new coating system with very different or opposite properties in a single coating layer. For examples, high hardness with low friction or high ductility. Another is high corrosion resistance with high electrical conductivity. For obtaining such kind of opposite properties at the same time, more than two phases should be formed in very small area especially, in nano-sized dimension. The most represent example of nanocomposite coating is DLC coating which consist of SP$_2$ and SP$_3$ phase and it has the properties of high hardness and low friction coefficient. Also this nanocomposite coating is very effective to prevent the corrosion process. The conventional coating with columnar structure has the original defect such as pinhole and pores and this could be a channel for the corrosion mass. So, they have a week properties for the corrosion or oxidation even they were made of ceramics. But, nano composite have very small and complex channel (grain boundaries) and also by a proper alloy design, corrosion or oxidation protective elements could be located inside grain boundary. The nanocomposite coating has the good corrosion and oxidation resistance [1-3].

Among the many important applications, the automobile is the most valuable market for the heat treatment and surface treatment industry. Because problems related to the Green house effect and demand on the high properties, automobile company have been tried to increase the fuel efficiency of the automobile by various method including high cost surface treatment that would not be considered ten years ago. The one of the effective methods is to make the low friction coating on engine parts. Theoretically, if we could reduce 50 % of the friction coefficient of engine parts, 10 % of fuel efficiency could be increased and this method is considered as the most effective one per cost [4]. In Korea, the Hyundai Motor Company (HMC) tried a DLC coating on tappet and received a good result. So, they transfer the DLC coating to other engine parts. But the engine operation conditions for the future automobile should be very harsh. It is thought that the property of DLC could not endure such conditions. For example, the temperature inside GDI engine is 650 °C in which DLC should be transformed into graphite and lost its good property [4]. So, they demand new coating materials with enhancing thermal stability. One of the possible and very powerful coating systems is suggested by Argon National Lab. and it is Mo-Cu-N system [5]. Since Mo-Cu is almost immiscible in liquid and solid state, Mo-Cu coating can be nanocomposite by a proper amount of Cu elements. Their friction coefficient in boundary region of the wear test in oil conditions is very good compared with the bare steel and even DLC coating in some specific condition. But because of the difficult of the alloying between immiscible phases, it is impossible to make such ternary coating in ordinary coating system.

Thus, ternary or binary coating in Mo-Cu system was preceded with using multiple sputtering targets [6].

At the case of using elemental targets, because the each element has its own sputtering yield, to get some composition we want we should control plasma power of each target and to get homogeneity we should rotate the sample. This resulted in the complicated equipment and process conditions. Also even if we use the alloying targets, if there have some inhomogenity, it is found that the coating layer has not the same composition with the target composition and the target has irregular surface erosion by the non-uniform sputtering [7]. So, it is very important to make the alloying target with very fine microstructure and homogeneous composition distribution to get the same composition on the coating with the target. In this study, the objects are to find out the possibility of the making the nanocomposite coating from the alloying target and to find out the possibility of the obtaining the coating with the same composition of target without dependent on the sputtering conditions.

EXPERIMENTAL DETAILS

By the conventional process, it is almost impossible to make the alloying target between Mo-Cu systems. In this study, powder metallurgy processes such as mechanical alloying and spark plasma sintering processes have been tried. The target making process was reported in other manuscript [8]. The alloying target has the disc shape with the size of 7.5 cm in diameter and 0.7 cm in thickness.

The coating was carried out with DC magnetron sputtering machine. The distance from target to sample is 5~7 cm. The coating was performed for 1 hr at room temperature with no bias. The DC power of the target was changed from 200 to 400 W. The chamber was evacuated by a turbo molecular pump to achieve the base pressure of 5×10^{-5} torr. Target was pre-cleaned until target voltage was the stabilized by the effect of sputtering of the contaminants on the target surface. During the target cleaning and stabilization of deposition condition, the shutter located before target was closed to prevent the deposition of the target materials on sample. The pre-cleaning was performed in Ar gas atmosphere and the reactive sputtering was in $Ar:N_2$ atmosphere with various gas ratio. The process pressure was varied form $2~10 \times 10^{-3}$ torr.

After coating the coating layer were examined by α-step, XRD, SEM & EDS, EPMA, AFM, TEM, nano-indenter, Scratch tester, wear tester. The coating thickness measured by α-step from scanning distance between the coated and bare substrate. The crystal structure was confirmed by XRD analyses with varying 2θ between 20 to 100°. Surface and cross sectional image of the films were observed by Field Emission Scanning Electron Microscopy (FE-SEM). The composition of coating and the surface composition uniformity were detected by Energy Dispersive Spectroscopy (EDS) and Electron Probe X-ray Microanalysis (EPMA). TEM was used to find out the formation of nanostructure and the position of Cu in the film. The surface hardness was measured by nano-indenter and the adhesion force was examined by scratch tester.

RESULTS AND DISCUSSION

Before the single target process, it had been surveyed on the properties of the coating by using two element targets of Mo and Cu. It has been tried to find the better composition in Mo-Cu system with higher mechanical properties. And then we made the alloying target with the same composition based on the results. The properties of the coating with the single alloying target were compared with those of the coating by dual target to find out the effectiveness of the alloying targets on the synthesis of the nanocomposite coating. Figure 1 shows the hardness, thickness, roughness of the coating prepared by dual targets. During the sputtering by double sputtering source with two element targets, the power of Mo source is fixed as 300 W and the power of Cu target was changed from 100 w to 500 W. SEM analyses with EDS showed that Cu composition could be obtained as 2-12 at. % in the coatings with dual targets. It was not easy to get some fixed Cu content and it is harder to get the high content of Cu in the MoN-Cu coating during the dual target process. This was because a resputtering of

Cu elements from the sample might be occurred at the high power sputtering process. Also the same reason the increase of the coating thickness was not observed definitely with the increase of Cu source power up to 400 W although Cu sputtering yield is higher than Mo, increase of Cu power should result in the increase of thickness of coating. The surface roughness was increased with the increased of Cu power. In two targets process, the hardness was the highest around 10 at. % Cu with the targets power of 300 W and 200 W for Mo and Cu, respectively. So, 10 at.% was selected as the composition to be compared.

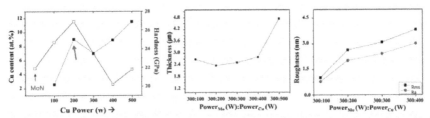

Figure 1. Changes of Cu percentage, hardness, thickness and roughness of the Mo-Cu coatings with the increase of the power for the Cu sputtering source during the double- target sputtering.

Mo-10 at.% Cu alloying target was made by MA and SPS process as mentioned previously and their microstructure was shown in Figure 2. Figures 2 are the SEM micrographs and EDS analyses data showing the microstructure and composition of the target. At the magnification of 20,000 in the SEM micrographs, the grain size of the target was less than 200 nm and according to EDS analyses, Mo and Cu elements was distributed well in the target but some segregation was found for Cu in the area that was considered as boundaries. However, the composition of the target was almost homogeneous in the submicron range and it was same with the as-milled powder. The detail description on target making process and the microstructure of the targets was reported in other manuscripts [8]. The microhardness of the target was around 650 Hv for 10 at. %Cu added target and 320 Hv for 50 at.% Cu added target and they had almost homogeneous hardness distribution within 5 % differences through the specimens.

Also, their hardness data were 50~100 % larger than those of commercially obtained ones with same composition, this was because of their fine microstructure resulting from the powder metallurgy process. It also found in our previous results that the target had enough mechanical properties such as strength, toughness and thermal stability. So, no problems were observed during the sputtering process.

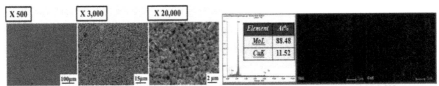

Figure 2. SEM and EDS micrographs of Mo-10 at.% Cu sample prepared by MA and SPS [8].

With the alloying target, sputtering was made for 30 minutes with the change of Ar:N₂ gas flow. The other variables were fixed as 300 W for sputtering power, 0 V for bias, room temperature. Figure 3 shows the SEM surface and close section images with increasing N₂ content. Only after 30 minutes coating, the thickness of all the coating was over 2 micrometer except the gas ration of 3:1. It was found that the deposition rate, the grain size and the surface roughness were decreased with the increase of N_2 amounts.

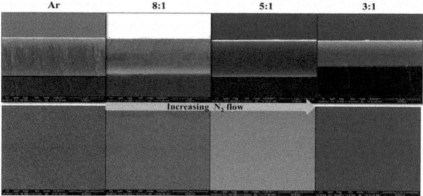

Figure 3. SEM surface images and cross section images of Mo-10 at.% Cu coatings with changes of Ar:N_2 gas flow.

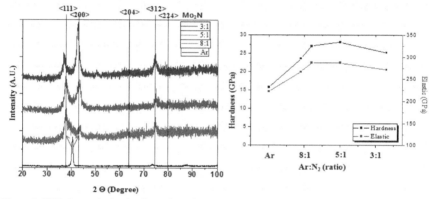

Figure 4. XRD data and microhardness data of Mo-10 at.% Cu coatings with different Ar:N_2 gas flow.

XRD analyses were examined to find out the change of the phases during sputtering process with increasing nitrogen content. Figure 4 shows the XRD data and the hardness data of the Mo-10 at.% Cu coatings with the increase of N_2 amount. XRD pattern shows that all the peaks were Mo_2N and Cu was not detected. This was because the Cu elements located at grain boundaries and could not be detected by XRD analyses [10]. Also, it was found in XRD data that the main peak changed form (111) to (200) with the increasing nitrogen content. The hardness that measured by nano indenter was increased with Ar:N_2 gas ratio. The highest hardness was observed at 5:1 with random orientation in XRD data. The hardness was decreased with further N_2 addition with the main peak of (200) in the XRD analyses. Thus, in this study, the gas ratio was fixed as 5:1 for Ar:N_2 because the coating had the highest hardness and a enough thickness compared with those prepared with other gas ratio.

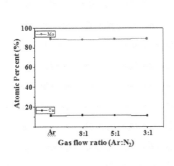
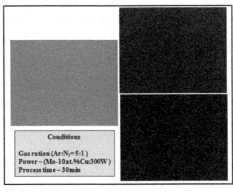

Figure 4A. EDS data of Mo-10 at.% Cu coatings with changes of Ar:N$_2$ gas flow.

The composition and distribution of Mo and Cu elements in the as-deposited coatings was examined by SEM and EDS methods. With change of N$_2$ content, the microstructure and hardness and even residual stress changes but the composition was almost same and Mo and Cu elements were homogeneously distributed through the coating layer. The composition of the coating was also confirmed by EPMA analyses as shown in table 1. If the composition of the coating was measured with nitrogen atom, Mo to N ratio was almost 2 to 1. This indicated the formation of Mo$_2$N phase. If the composition was measured without nitrogen atom, it was nearly same with that of target. Also, according to the GDOES data, the composition of the coating was very stable and consistent through the entire coating layer. In this study, it is found that the alloying target with the fine microstructure is very effective to make the coating with the same composition with the target.

Table 1. EPMA data on the composition of the coating layer prepared with Mo-10 at.%Cu target.

Target Composition (at%)		N(at%)	Mo(at%)	Cu(at%)
Mo-10 Cu	w N	26.24	66.90	6.86
Mo-10 Cu	w/o N	0	90.93	9.07

To find out the advantage of the process using alloying target respect to the co sputtering process with two elemental targets, the properties of the coatings prepared by single alloying target were compared with those prepared with two elemental targets. Figure 5 shows the SEM cross section image and surface roughness data of the coatings by two targets and single alloying target. As shown in the figure 5, the microstructure of the coating from two targets was columnar structure. But the coating by alloying target had the perfect featureless structure even the coating was prepared at room temperature. Interestingly, the coating layer by alloying target used only 300 W as a power for sputtering target was thicker than that of two targets with 300 W for Mo target and 200 W for Cu target. Also the surface roughness of the coating from alloying target was three times better than that of two targets.

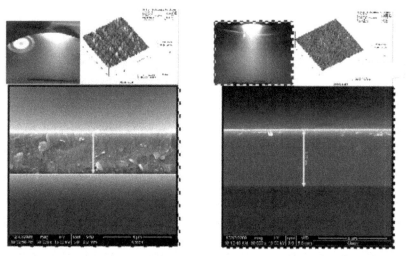

Figure 5. SEM cross section images and surface roughness data of the coatings prepares by two elemental targets and the single alloying target, respectively.

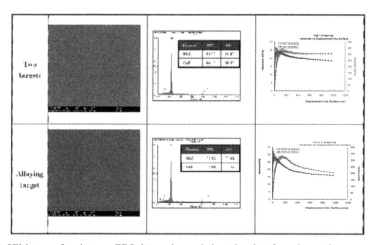

Figure 6. SEM top surface images, EDS data and nano indentation data from the coatings prepared by two elemental targets and single alloying target, respectively.

Figure 6 shows the SEM top surface images with EDS data, and nano indentation data from the coatings made by two targets and the alloying target, respectively. The coating prepared by two targets had the grain size around 50nm. The coating from alloying target has the finer microstructure so that it was not measureable in SEM micrographs. The finer microstructure of the coating from alloying target was resulted in 20 % higher hardness (28-29 GPa) compared with that (24-25 GPa) of the coating with the same composition and prepared from two targets. The coating from alloying target has almost the same composition with the target composition and moreover, it was easily prepared by just sputtering with proper power and without changing any other variables. But with two elemental targets, as mentioned previously, it was not easy to get the composition that would be intended to be obtained (10 at.% Cu) and the final composition was 8.9 at. % Cu. Another good properties of the coating produced by alloying target was the low E (that is elastic modulus) value at the same hardness, compared with that from co-sputtering process. It meant they had high H/E value that is index of the high toughness of the coating. So, it is thought that the mechanical properties of the coating prepared from alloying target should be better than that from co-sputtering.

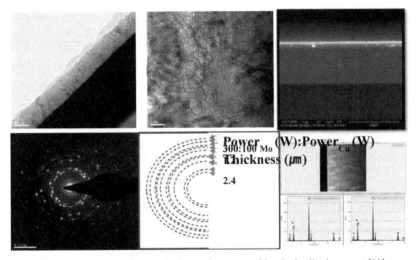

Figure 7. TEM micrographs of Mo-10 at.% Cu coating prepared by single alloying target [11].

Figure 7 shows the TEM data of Mo-10 at.% Cu coating prepared by sputtering from single alloying target. According to SEM data, almost featureless microstructure was observed as shown in far left picture. But TEM analysis shows the nano columnar structure with the width of 10 nm. Inside the nano columnar layer, the grain size was very small and less than 5 nm as measured by confirming the domain size with the same direction of atomic layer in high resolution TEM micrograph. The fine polycrystalline microstructure in the coating resulted in a ring patterns by a Selective area patterns in HRTEM. Diffraction patterns showed the γ-Mo_2N peaks and some trace of Cu peaks. But it is almost impossible to detect Cu that would not exist as the second phase in the grain but seems to be in the vicinity of the grain boundaries. However it was detected that Cu elements was distributed homogeneously in the coating layer according to the EDS data by HRTEM analyses.

CONCLUSIONS

In this study, the coatings prepared by sputtering with an alloying target were compared with those prepared with two elemental targets. By using alloying targets, nanocomposite coatings that have the same compositions with the targets and nano-sized microstructure could be obtained easily by reactive sputtering process at room temperature. This coating has the better properties compared with those prepared by co sputtering process with two and more elemental targets. The microstructure and the surface roughness of the coatings from the single alloying target were finer compared with those of the coatings prepared by co sputtering process. Thus, the coating from single alloying target had the hardness 20 % higher than that prepared from two elemental targets. With the 10 at.% Cu composition, the coating has the hardness between 28~29 GPa. According to TEM analyses, the nano-columnar structure was observed and the size of the grain inside the nano-columnar structure was less than 5 nm. The trace of Cu elements could be detected by HRTEM analyses but it was impossible to find out the exact location of Cu elements in this study.

ACKNOWLEDGEMENT

This research was supported by a grant from the Fundamental R&D Program for Core Technology of Materials funded by the Ministry of Knowledge Economy, Republic of Korea.

REFERENCES
[1] S. Veprek, R.F. Zhang, M.G.J. Veprek-Heijman, S.H. Sheng, A.S. Argon, Superhard Nanocomposite: Origin of Hardness Enhancement, Surface & Coatings Technology, 204, 1898-1905 (2010)
[2] J. Musil, P. Zeman, H. Hruby, P.H. Mayrhofer, ZrN/Cu Nanocomposite Film – A novel Superhard material, Surface & Coatings Technology, 120-121, 179-183 (1999)
[3] A. Leyland, A. Matthews, Design Criteria for Wear-Resistant Nanostructured and Glassy-Metal Coatings, Surface & Coatings Technology, 177-178, 317-324 (2004)
[4] W.S. Cho, Strategy for Green Automobile Technology, Plenary Talk, Seventh Asian –European International Conference on Plasma Surface Engineering, Busan, Korea, (2009)
[5] A. OZturk, K.V. Ezirmik, K. Kazmali, M. Urgen, O.L. Eryimaz, A. Erdemir, Comparative Tribological Behaviors of TiN, CrN, and MoN-Cu Nanocomposite Coatings, Tribology International, 41, 49-59 (2008)
[6] M.C. Joseph, C. Tsotsos, M.A. Baker, P.J. Kench, C. Rebholz, A. Matthews, A. Leyland, Characterisation and Tribological Evaluation of Nitrogen-Containing Molybdenum-Copper PVD Metallic Nanocomposite Films, Surface & Coatings Technology, 190, 345-356 (2005)
[7] W. Zhenxia, P. Jishen, Z jiping, W. Wenmin, Localization of the constituent depth profile near Al-Sn alloy Surface, Vacuum, 46, 1271-1273 (1995)
[8] H. C. Lee, B.K. Shin, K.I. Moon, A Study on the Microstructure and Properties of Mo-Cu alloys Prepared by Mechanical Alloying and Spark Plasma Sintering Process, in Preparation, J of alloys and compound, (2011)

CUSTOMIZED COATING SYSTEMS FOR PRODUCTS WITH ADDED VALUE FROM DEVELOPMENT TO HIGH VOLUME PRODUCTION

Dr.-Ing. T. Hosenfeldt, Dr.-Ing. Y. Musayev, Dipl.-Ing. Edgar Schulz
Schaeffler Technologies AG & Co. KG ,
Industriestr. 1-3, 91074 Herzogenaurach, Germany.

ABSTRACT
Modern components and systems for automotive and industrial applications have to meet various requirements in multiple technical fields. Apart from properties that affect the part itself – like geometry, stiffness, weight or rigidity – the surface properties must be adjusted to the growing environmental requirements. Therefore coatings are increasingly applied to reduce the friction losses of car components, improve fuel efficiency and reduce CO_2-emissions. This article describes how to use surface technology as a modern design element for components and systems to enable the demanding requirements on market leading automotive and industrial products. Therefore Schaeffler has developed and established a coating tool box for customized surfaces to deliver the right solutions for all that needs and requests with the corresponding coating system enabled by PVD-/ PACVD-, spraying or electrochemical technology. For innovative products it is extremely important that coatings are considered as design elements and integrated in the product development process at a very early stage. In this article tribological coatings are viewed within a holistic and design-oriented context. The latest developments of amorphous carbon coatings, their characteristics as well as the technical and economical effects of their use in combustion engines and industrial bearing applications are described. The presented Triondur® amorphous carbon based coating systems (a-C:H; a-C:H:Me; a-C:H:X and ta-C) are excellent examples for customized tribological systems like bucket tappets or roller bearings. These carbon coatings offer the following advantages: super low friction with highest wear resistance, customized surface energy, optimized wetability and interaction with formulated engine oils and low adhesion to the counterpart.
A close collaboration between designers and surface engineers is required in the future. Schaeffler delivers around 75 million high-quality PVD- and (PA)CVD-coated components every year that enable outstanding applications, preserve resources and meet increasing customer requirements.

1 MOTIVATION

Today CO_2 reduction is often used as a synonym for friction reduction in the automotive industry. Above all, legislative guidelines, for example these of the European Union, to regulate CO_2 emissions and fleet consumption of automobile manufacturers are the driving forces for the improvement of energy efficiency and durability of modern gasoline engines[1]. This means, that it is getting more and more important to reduce friction and wear in highly stressed contacts of combustion engines. Therefore, the application of amorphous carbon coatings, used for example in the valve train, is in demand. In such tribological systems the part, counterpart, intermediate and environment interact each with another in a manifold manner (Figure 1)[2].

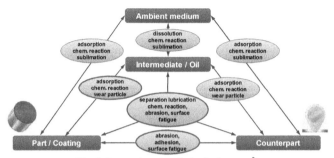

Fig. 1: Interactions in tribological system[2].

In the contact area between part and counter part elementary friction and wear mechanisms such as adhesion, abrasion, surface fatigue, deformation, elastic hysteresis and damping, tribochemical reactions etc. may occur. These mechanisms are overlapping in the temporal and spatial undetectable real material contact area. In addition highly formulated lubricants, which are mainly used with highly stressed engine elements, exert influence on tribochemical reactions inside and outside of the tribological contact[2].

Due to these various reciprocal effects the behavior of such tribological systems becomes very complex, which means that the theoretical prediction of friction and wear is hardly possible at present. Therefore during development and/or optimization of each tribological system, always experimental investigations mostly based on empirical knowledge are conducted. Thereby normally the focus is only on one machine part or even exclusively on the coating. Ideally such investigations are carried out on test rigs, which reproduce the real system as close as possible, for example on a driven cylinder-head test rig, to investigate the behavior of the machine part respectively to the coating under real boundary conditions. However, such experiments are associated with a high temporal and financial effort. This often leads to the application of abstracted, cheap and time effective model tests or component tests like the ball-on-disk tribometer. However the transferability of these tests to the application is restricted. This indicates the need for a suitable method to predict the friction and wear behavior of tribological systems containing tribological coatings. With this development times and the amount of expensive tribological tests can be reduced.

2 AMORPHOUS CARBON COATINGS

Since the mentioned interactions take place mainly at the surface of contacted components, friction and wear can be influenced by surface modification like structuring and coating. By coating the ground part for example, a further element is added to the structure of the tribological system (Figure 2)[7]. These surface modifications have to be adapted to the respective application and should be taken into account during the design phase of the tribological system, so the coating should be seen as a design element.

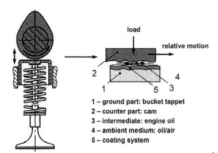

Fig. 2: The five elements of a tribological system[7].

1 – ground part: bucket tappet
2 – counter part: cam
3 – intermediate: engine oil
4 – ambient medium: oil/air
5 – coating system

Amorphous carbon coatings are in particular suitable for this purpose since its properties can be adjusted for different requirements. These characteristics are founded on its bonding structure and hydrogen content. Amorphous carbon coatings consist of notably stable diamond like tetrahedral sp^3-bonds, which ensure a high wear resistance, and graphitic sp^2-bonds, that enable a slight slipping between the atomic hexagonal structured planes[8, 9, 19, 20]. Regarding its structure and its properties amorphous carbon coatings can be located in a triangle between graphite, polymer and diamond.

Depending on the manufacturing process, the coating parameters and the exact composition, it is possible to adjust the properties of the coatings up to the edges of the triangle. In addition the chemical structure and polarity of the coating, thus the reactivity with surface active additives can be influenced by doping the coating with metals (Cr, W, Ti, ...) or nonmetals (F, O, Mo, Si, ...)[8]. Schaeffler has developed a coating toolbox with tailored amorphous carbon based coating systems (a-C:H; a-C:H:Me; a-C:H:X and ta-C) as a modular design element for customized tribological systems with low friction and highest life time.

The advantages of amorphous carbon coatings under dry conditions are the low friction coefficient of about 0.1 to 0.2 in combination with high wear resistance[9]. In lubricated tribological systems working under mixed friction conditions the performance of amorphous carbon coatings depends on the interaction between all elements of the tribological system. Concerning its additive composition, current engine oils are optimized for steel/steel contacts. Accordingly the chemical and physical behavior of additives, like anti-wear agents, extreme pressure agents and friction modifiers, in contact with steel is well known. If however engine parts are coated with amorphous carbon coatings, the chemical behavior of one element changes and therefore the behavior of the whole tribological system. The reason for this is that these additives (e.g. Mo-DTC, Zn-DTP, ...) cannot completely take effect on the inert and non-polar carbon surface. Through the changed reaction conditions even undesirable effects may occur leading to an increase of friction and wear[10, 11, 19, 20].

3 PREDICTION OF FRICTION AND WEAR

The goal during an application-based and market-oriented development of tribological coatings for reduction of CO_2 emissions is to provide a coated system solution respectively coated components with a maximum of customer benefit as quick and cost-efficient as possible. To reduce development time and coating cost, lubricant and counter part need to be adjusted purposefully to the particular application. This means for the system bucket tappet/ camshaft, that with a given component – for example camshaft, lubricant or coating – a recommendation for the design of the remaining components can be given.

One possibility to achieve this goal is the isolated determination of single effects as published in[10 – 17]. Thereby the single effects are observed under fixed boundary conditions realized by abstraction and simplification of the tribological system in order to handle the complexity of the tribological processes. Thus in[10] the reactivity of the anti-wear agent Zn-DTP could be verified on an amorphous carbon coating containing hydrogen by the use of a ball-on-disk tribometer with rotating disc and an additional XPS analysis. However the adhesion strength of the resulting tribofilm was considerably weaker than on a steel surface. As a result the wear reducing effect of Zn-DTP cannot occur and on the contrary hard wear promoting particles are brought into the tribological contact by delamination of the reaction layer. Moreover, the formation of a reaction layer by Zn-DTP requires energy, which leads to an increase of the friction coefficient up to level of the steel/steel contact[10]. However, it is possible to achieve a significant friction reduction in the range of fluid friction with hydrogen free carbon coatings when the lubricant is adjusted to the carbon surface. In[12], it was found that special friction modifiers are especially effective with tetrahedral hydrogen free amorphous carbon coatings (ta-C). In a ball-on-disc tribometer with oscillating ball, ta-C coated disc and a GMO doped base oil friction coefficients below 0.02 were reached. This extreme low friction is explained by the hydroxylation of the ta-C surface leading to low shear forces in the interface between coating and lubricant[12]. With amorphous carbon coatings containing hydrogen such a drastic friction reduction was not yet reached. In[11] various amorphous carbon coatings were investigated with a lubricated block-on-ring test lubricated with a Mo-DTC friction modifier doped base oil. Here, it was found that Mo-DTC decomposes in the tribological contact under generation of MoS_2 and MoO_3. Although the friction coefficient was reduced with most amorphous carbon coatings, a rise in the wear amount was observed. MoO_3 seems to promote the oxidation of the amorphous carbon coatings[11].

These described investigations were often carried out with model tests like pin on disk, which can be carried out quite easy, but the results are often far away from the real application. On the other hand, aggregate and field tests have a much better transferability to the application, but it is to be expected that these test are more expensive and time-consuming.

Just in such tribological systems like the contact between cam and coated bucket tappet with lubricants containing additives, it is quite possible that results out of different categories can be contradictory. This was proven in[18] by carrying out experiments with different amorphous carbon coatings on a ball-on-disc tribometer and a driven cylinder head rig, whereby no correlation between the results could be found. So based only on model tests, it is not possible to fathom out all effects and superimpositions for a transfer to the real system cam/bucket tappet. Up to now this transferability can only be ensured when aggregate tests or even field tests are applied, whereby the financial and temporal efforts become disproportionately high. But applying statistical methods with learning capabilities, the experimental effort can be reduced significantly. Thereby it is possible to determine expected output variables from given input variables. For example artificial neural networks allow a quite good prediction of friction and wear on the basis of a large amount of experimental data.

4 DATA COLLECTION

A predictive modeling requires the availability of a sufficient large amount of empirical test data. For this database already existing test data from a driven cylinder head test rig were pulled up. Furthermore, a multitude of experiments was carried out by a systematic variation of the explanatory variables with a ball-on-disk tribometer. As already mentioned, there is an obvious discrepancy between the results of these two tests. Therefore, a novel valve train test rig, which should close the gap between the model test and the aggregate test, was used. It is designed as a single valve test rig and reproduces the system cam/bucket tappet as realistic as possible. Thereby a genuine camshaft section with the associated bucket tappet and valve spring is used. This valve train model test rig provides the opportunity to determine the friction and normal forces appearing in the tribological contact separately.

The used database generated with all these different tribological methods includes a total amount of 262 experiments (NDO).

In case of the tribological contact between cam and coated bucket tappet, the output variable is the friction coefficient. The input variables are the type of coating, hardness, Young´s modulus and the surface roughness. Thereby aside from uncoated bucket tappets and a nitride chromium hard coating, six different amorphous carbon coatings were investigated. The used amorphous carbon coatings differ in their hydrogen contend and their doping elements. The lubricant additives, their concentration, the base oil and its viscosity were added to the database as well. Here, PAO and mineral base oils as well as the additives Mo-DTP, Zn-DTP and GMO were taken into account. Furthermore, the stress parameters of the tribological experiment were varied as input variables, so does relative velocity, oil temperature and contact pressure. The reproducibility of the friction measurements was determined by repeating a measurement several times at the same conditions. These results show that the relative variance of these measurements is in a range of about 5% of the friction value.

APPLICATIONS

With the holistic competence in surface technology, we offer prediction and optimization of tribological systems, design of the tailored coating system and its production process for components with added value for the customer and society for automotive as well as industrial applications.

The valve train is one of the main elements of combustion engines and is closely involved in the friction losses of the combustion engine[3]. The valve train can assume several designs depending on the position of the camshaft and design of the cam follower. Above all, the mechanical bucket tappet and the roller finger follower are at present the most common valve train systems[5]. With the roller finger follower, the contact between cam and follower is designed as a rolling contact with low friction losses. However, the realization of a rolling contact is considerably more expensive[3,6]. This is reflected in the cost benefit ratio as well (Figure 3).

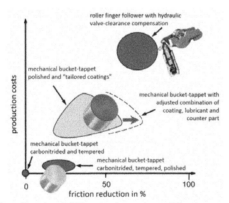

Fig. 3: Cost-benefit ratio of different valve train modifications according to [6]

Here the basis with 0% friction reduction and 100% costs is represented by the carbonitrided and tempered mechanical bucket tappet. A friction reduction of about 25% can be achieved by an optimized surface treatment of the standard bucket tappet. An additional reduction of friction down to the low level of a roller finger follower is possible by the application of a tailored coating (Figure 3)[6].

With an adjustment of all components, for example the best coating for the respective lubricant or the respective cam and cam structure, it is possible to reduce the friction level down to 50%, what means lower CO_2 emissions of 1% to 2%.

Legislative guidelines will increasingly influence this cost-benefit approach[1]. With this the greater expense for coating engine parts can be amortized by the reduction of CO_2 emission.

Energy efficiency is getting also more and more important in industrial applications and the operational areas are facing tougher challenges. Therefore Triondur® coating systems has gained attention as a modular and successful design element in product development to reduce the friction losses and enhance the operational area respectively the life time of bearings. With a tailored DLC coating system, it is possible to reduce the friction in barrel rollers under poor lubrication, nearly independent from the load of about 40%.

In other words, such coating systems enable the same low friction under poor lubrication than standard bearings under good lubrication (Figure 4).

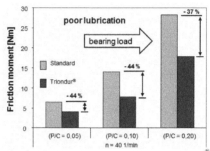

Fig. 4: Friction reduction in barrel roller bearings with Triondur® coating systems under poor lubrication.

5 CONCLUSION AND OUTLOOK

A close collaboration between designers and surface engineers is required in the future. With the Global Technology Network, Schaeffler is combining its existing local expertise in the regions with the knowledge and innovative force of its experts all over the world. This will strengthen the local expertise and bringing engineering and service knowledge even closer to its customers through the introduction of Technology Centers and collaboration in global expert networks of industry and science. Customers all over the world benefit from this and receive innovative, customized solutions of the highest quality.

Schaeffler delivers around 75 million high-quality PVD- and (PA)CVD-coated components every year that enable outstanding applications, preserve resources and meet demanding customer requirements. In order to reduce the temporal and financial effort for tribological experiments, a prediction of the friction behavior of lubricated tribological systems containing amorphous carbon coatings should be realized. This prediction requires the availability of a sufficiently large database which contains case studies.

ACKNOWLEDGMENT

We acknowledge the funding of our work on prediction of friction and wear in the research project "PEGASUS" by the Federal Ministry of Economics and Technology (BMWi).

References

[1] Regulation (EC) Nr. 443/2009 of the European Parliament and of the Council of 23 April 2009.

[2] H. Czichos, K.-H. Habig: Tribologie Handbuch, Wiesbaden: Vieweg, 2010.

[3] P. Gutzmer, MTZ, 68 (2007), 243.

[4] F. Koch, U. Geiger, tribological symposium of GfT and DGMK, Göttingen, 05. 06.11.1996.

[5] M. Lechner, R. Kirschner, in: R. van Basshuysen, F. Schäfer (Eds.): Handbuch Verbrennungsmotor. Grundlagen, Komponenten, Systeme, Perspektiven. Braunschweig/Wiesbaden: Vieweg, 2002.

[6] A. Ihlemann, G. Eggerath, T. Hosenfeldt, U. Geiger, 14th Aachener colloquium vehicle and motor technology, Aachen, 04.-06.10.2005.

[7] Y. Musayev, Verbesserung der tribologischen Eigenschaften von Stahl/Stahl-Gleitpaarungen für Präzisionsbauteile durch Diffusionschromierung im Vakuum, 2001.

[8] VDI Guideline 2840: Kohlenstoffschichten – Grundlagen, Schichttypen und Eigenschaften. Düsseldorf: VDI, 2005.

[9] J. Robertson, Material Science and Engineering R37 (2002) 129-281.

[10] S. Equey, S. Roos, U. Mueller, R. Hauert, N. D. Spencer, R. Crockett, Wear 264 (2008) 316–321.

[11] T. Shinyoshi, Y. Fuwa, Y. Ozaki, Spring Academic Lectures of the Automobile Technology (2007).

[12] M. Kano, Tribology Letters, Vol. 18, No. 2, (2005) 245-251.

[13] A. Erdemir, Tribology International 37 (2004) 1005–1012.

[14] T. Haque, A. Morina, A. Neville Surface & Coatings Technology 204 (2010) 4001-4011.

[15] T. Haquea, A. Morina, A. Neville , R. Kapadiab, S. Arrowsmith Wear 266 (2009) 147–157.

[16] Ksenija Topolovec-Miklozic, Frances Lockwood , Hugh Spikes Wear 265 (2008) 1893–1901.

[17] B. Podgornik, J. Vizintin, Surface & Coatings Technology 200 (2005) 1982 – 1989.

[18] E. Schulz, Y. Musayev, S. Tremmel, T. Hosenfeldt, S. Wartzack, H. Meerkamm, Magazin für Oberflächentechnik 65 (2011) Nr. 1-2, S. 18-21.

[19] C. Donnet, A. Erdemir (ed.), Tribology of Diamondlike Carbon Films: Fundamentals and Applications, Springer, New York, 2008.

[20] A. Erdemir, C. Donnet, Tribology of Diamondlike Carbon Films: Current Status and Future Prospects (topical review)," Journal of Physics D: Applied Physics, 39, (2006) 311-327.

A STUDY ON THE IMPROVEMENT OF THE SERVICE LIFE OF SHAFT-BUSHING TRIBO-SYSTEMS BY PLASMA SULFUR NITROCARBURING PROCESS

Kyoung Il Moon[1], Hyun Jun Park[1], Hyoung Jun Kim[2], Jin Uk Kim[2], Cheol Wong Byun[2]
[1] Korea Institute of Industrial Technology, Incheon Division, Heat & Surface Technology Center, Incheon 406-840, Korea, kimoon@kitech.re.kr
[2] Daekeum GEOWELL, Research Center, Incheon 402-061, Korea, hjkim@daekeum.co.kr

ABSTRACT

To improve the service life of the shaft-bushing tribo system, the bushing was treated by plasma sulfur nitrocarburizing process. Before a plasma sulfur nitrocarburizing, the specimens were treated by plasma nitriding. As a result, a compound layer was formed with thickness about 20 μm. After then, by introducing a plasma sulfur nitrocarburizing process, a very fine and porous FeS layer with the grain size less than 100 nm was formed on the top of the compound layer. This fine and porous FeS layer formed on the samples led to the improved friction coefficient and resulted in the better wear resistance during the ball on disk test in dry conditions. Moreover in oil conditions, the fine and porous structure in the FeS layer could act as an oil pocket, so the wear properties would be much improved. So, it is considered that this fine and porous FeS layer could replace the PTFE and MoS_2 coating that have been commercially used in busing parts. So, in this study, the endurance tests have been tired on the commercialized bushings and the plasma sulfur nitrocarburizing bushing by a rig test machine that was designed and made by our research team to estimate the durability of the plasma sulfur nitriding bushing. Although the S bushing had higher friction coefficient and resulting higher operating grease temperature, they had a longer expected service time that was determined on the point where a rapid increase of the friction coefficient occurred. This must be FeS layer on the top of the compound layer had enough endurance during the operation condition for the bushing. So it is considered that the S coating could be used for the busing instead of M coating.

1. INTRODUCTION

Sliding bearing in a low speed and high load condition is one of the most important parts in the front hinge parts of excavators [1-2]. The sliding contact between a shaft and a bushing is known to have a microscopic irregularity. Thus the damage of the shaft and the bush are formed irregularly. Especially, the localized plastic deformation of the bushing results in a severe damages. In many cases, the damage is accelerated by the fragments of part or dust between the bushing and the shaft. The well known damages of the bush are summarized as abrasive wear, fatigue cracking, cavitation, and adhesive wear [3-4]. The accumulation of microscopic damage in the parts leads to severe wear, sliding noise and final macro-fracture. Thus, the shaft and the bushing need to have the properties of high strength, high wear resistance and good lubrication [5]. Generally, they are surface treated such as carburizing, nitriding and then low friction treated like PTFE (Polytetrafluorethylene) coating or MoS_2 coating [6-7]. The PTFE and MoS_2 coated bushings are commercial products. But such processes are two steps process, that is, they are treated to be nitride in one chamber and then subsequent coated in another chamber. MoS_2 and PTFE coatings are well-known solid lubricants. So, they could be easy sliding surface at the surface of bushing parts. Also the hardened layer prepared by nitriding process could prevent the damage such as fatigue fracture during the repeated solid contact between bush and shaft if the nitriding layer has enough hardened depth. But such two steps process increases the cost of production. Also the processes related to the salt bath nitriding and MoS_2 coating is very detrimental to environment. So, if the required properties could be gained by one step process and clean process, the production cost would be cut off and the working environment could be improved.

In this study, the surface of bushing has been treated by plasma sulfur nitrocarburizing process (carbon added plasma sulfur nitrocarburizing process). According to our previous results [8], the addition of carbon source in plasma sulfur nitriding process was resulted in the much fine surface structure and this induced much improved wear properties. Also we prepared a commercialized bushing which were treated by salt bath nitriding and MoS_2 coating and a bushing which were treated by carburizing and Sulf BT process which is well-known process of HEF Company [9]. A set of endurance tests on the heat treated bushings were carried out by a rig tester in order to find out the life time of each bushing. Based on the rig tests, the possibility of using the bushing that was prepared by plasma sulfur nitrocarburizing process as the parts of the excavator has been discussed.

2. EXPERIMENTAL DETAILS

2.1. Materials and heat treatments

The bush was generally made with SCM440 steels (KS standards) with a chemical composition (wt.%) shown in Table 1. The bushing had a cylindrical-shaped with an inner diameter of 71 mm and an outer diameter of 86 mm and an axial length of 60 mm. Disk shaped SCM440 samples were also prepared with a diameter of 30 mm and 10 mm thickness. The disc samples were used for the measurement of the microstructure and the mechanical properties of the heat treated sample. All the samples were heat treated in the vacuum chamber and oil quenched and tempered at high temperature around 550°C. After heat treatment of quenching and tempering, the bushing and steel sample had the microstructure characterized by a fine tempered Martensite. Before the nitriding, they were cleaned ultrasonically in ethylene bath and dried before placement into a plasma nitriding chamber.

Table 1. Chemical composition of the specimen

Material	Composition (Wt.%)								
	C	Si	Mn	P	S	Ni	Cr	Mo	Cu
SCM440	0.38~	0.15~	0.60~	<	<	-	0.90~	0.5~	-
(AISI4140)	0.18	0.35	0.85	0.030	0.030		1.20	0.30	

Then samples were loaded in the plasma nitriding equipment shown in Fig. 1. The characteristic feature of this machine is that there is a subsidiary cathode surrounding the working zone. By this subsidiary plasma generator, nitriding is possible without plasma directly charging on the samples. Except many advantages such as high productivity and easy charges of workloads with different geometry and size, and avoiding the common problems associated with conventional ion nitriding [10, 11], a rapid nitriding is possible in this chamber with a bias cathode and a subsidiary cathode. By plasma generation on both substrate cathode and subsidiary cathode, the surface density of ion could be doubled and this induced a rapid nitriding. That is, the compound layer and hardened layer formed much faster than those in the process only with the substrate cathode.

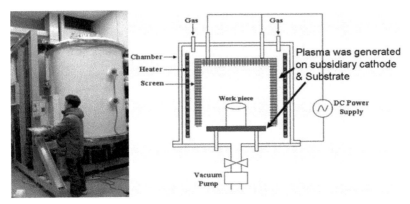

Figure 1. Figure of nitriding chamber designed by KITECH and the Schematic diagram of the subsidiary plasma nitriding process

Table 2. The process conditions of the plasma sulfur nitrocarburizing

	Temp. ($^\circ$C)	Pressure (mbar)	Time (min.)	Voltage	Gas (sccm)				
					Ar	H$_2$	N$_2$	N$_2$-H$_2$S	C$_3$H$_8$
Pre-treatment	550	1.5	30	600~700	500	500			
Nitrocarburizing	550	2.5	120	500	-	2000	2000		100
Sulfur nitrocarburizing	550	2.5	120	500		2000	2000	100~700	100

In this study, plasma sulfur nitriding was performed with bias and subsidiary cathodes connected to the DC pulse power supply with the capacities of 50 kW & 20 kW, respectively. Table 2 shows the process conditions of this study. The base pressure for sulfur nitriding was 6×10^{-3} torr that was attained by rotary pump and booster pump, and then the temperature was increased to 550 °C. Before main process, the sample was pre-cleaned with Ar/H$_2$ plasma at the pressure of 1.5mbar for 30 minutes during the heating process. The bias voltage of the pre-cleaning was – 600~700V. The details on the plasma sulfur nitriding were described in other manuscripts [12]. Base on the conditions for the best properties could be obtained; main process of plasma sulfur nitrocarburizing was decided. During the plasma sulfur nitrocarburizing, mixed gases were introduced with the ratio of N$_2$, H$_2$ = 1:1. The nitrogen gas and the hydrogen gas were fixed as 2000 SCCM, respectively. H$_2$S gas was 5 % mixed with N$_2$ gas and this 5 % H$_2$S/N$_2$ mixed gas was added from 700 SCCM. Thus, the amount of H$_2$S gas was added about 35 SCCM. The process pressure was 2.5mbar. In additions, 2% C$_3$H$_8$ gas of the total gas was added to be a nitrocarburizing. With subsidiary plasmas, plasma nitrocarburizing was performed for 2 hours with only N$_2$ and H$_2$ and 2% C$_3$H$_8$ gas. This process led to the formation of compound layer with 20 μm thickness and nitriding layer with 400 μm on the surface of the bushing. After then, plasma sulfur nitrocarburizing was performed for 2 hours as prescribed above. As a results, over 5-10 μm FeS layer was additionally formed at the top of the compound layer.

Comparing with the bushing treated with plasma sulfur nitriding, two commercial bushings were prepared for the rig endurance tests. So the test bushings were prepared as follows.
1) Salt bath nitriding + MoS_2 (M bushing) – commercialized one
2) Carburizing + Sulf BT [9] (B bushing)
3) Nitriding + Plasma sulfur nitrocarburizing (S bushing)

The shaft pin used for the counter parts in the tribological tests were produced with SCM440 and induction heated, oil quenched and then tempered at low temperature with a resulting hardened depth of 1.5 mm.

2.2. Characterization

Microstructures of heat treated bushings were observed by optical microscope (Z16APO) and scanning electron microscope (SEM, HITACH S-4300). The surface hardness and case depth were measured with micro Vickers hardness tester (Future Tech FM-7). Surface roughness was observed with atomic force microscope (AFM, NS4A). X-ray diffraction (XRD, Rigaku RAD-3C) was used for phase analysis of sulfur nitriding layer. Chemical composition was analyzed using electron probe micro analyzer (EPMA, Shimadzu EPMA-1400). High resolution TEM (Hitachi, 300 kV) was used to investigated on the cross sectional image of the surface area that would be formed during plasma sulfur nitrocarburizing.

Figure 2. The feature of the rig test machine.

The endurance tests on the heat treated bushings were performed using a rig test machine that was designed and made by our research team as shown in Fig. 2. The rig test machine was specifically manufactured so as to accommodate bushing and shaft assemblies up to a nominal coupling diameter of 71mm, with a maximum radial load capability of 300 kN. Generally the bushing in excavator had a pressure of 500 kgf/cm^2 and an angular velocity of 1-2 m/min as operation conditions. So in this study, the machine was operated in alternating rotation, with a span angle of 100 degrees, at an angular speed of 1.24 m/min. For the acceleration of the tests, the test pressure was decided as 750 kgf/cm^2 (181KN). This value was 1.5 times higher than that of operation condition. Grease for the lubrication was provided once for all before starting the endurance tests, which was carried out at room temperature. The rig test machine was installed with a torque meter, a thermocouple and a load cell, in order to provide the real time measurements of the torque generated by friction between pin and bushing, the grease temperature and the applied radial load. Endurance testing was continued until the rapid rise of a friction coefficient or grease temperature that meant the severe wear between the pin and the bushing.

3. Results and discussion

3.1 Microstructure analysis on the heat treated bushings

Figure 3. The optical micrograph of surface and cross-section of SCM440 nitrided samples in gas ratio of H$_2$S : C$_3$H$_8$ = 700 : 100 (unit, SCCM) [8]

In our previous study [8], it had been tried to make plasma sulfur nitrocarburizing on the surface of compound layer to replace MoS$_2$ coating process that was performed for the commercialized bushing. By introduce the C$_3$H$_8$ with a gas ratio of H$_2$S (it could be represented as a plasma sulfur nitrocarburizing), the microstructure was decreased to nano sized range less than 100 nm Fig. 3. Especially in the surface of the specimen prepared by 700 SCCM H$_2$S and 100 SCCM C$_3$H$_8$ gases, large amount of very fine pores was formed as shown in Fig. 3. The porous structure was not definitely observed in SEM cross section image because the pore was too small below 100 nm as shown in left picture of Fig. 3. But according to the observation on the cross sectional microstructure in the sulfur nitriding layer by HRTEM investigation as shown in Fig 4, this porous structure was confined surface layer less than 500 nm.

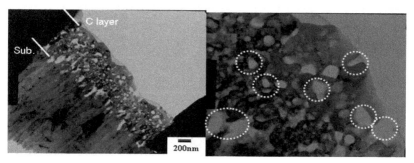

Figure 4. High resolution TEM micrographs showing the cross section microstructure of the specimen prepared by plasma sulfur nitrocarburizing in H_2S and C_3H_8 gases.

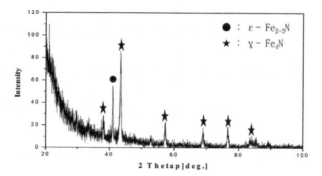

Figure 5. XRD analysis on the sample prepared by plasma sulfur nitrocarburizing in H_2S and C_3H_8 gases.

Figure 5 is the XRD analysis on the surface area on the plasma sulfur nitrocarburized SCM440 steel. The XRD analysis showed that the compound layer of the plasma sulfur nitocarburized specimen consisted of ε -$Fe_{2-3}N$ and γ -Fe_4N phases but the peak of FeS was not detected by XRD analysis. It is thought this was because the FeS layer was so thin to be analyzed by XRD or S was not formed the FeS during the plasma sulfur nitrocarburizing process. Also there were some reports that the FeS peak was not effectively detected by XRD because the Peaks for FeS phase were overlapped with those of other peaks [12, 13].

Figure 6. XPS plasma sulfur three different

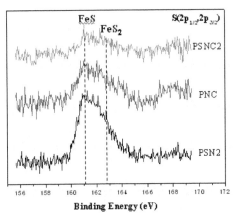

analyses of the samples nitriding treated in conditions.

However, analyses in Fig. could be levels of 161, So, it is peak was plasma sulfur

according to the XPS 6, FeS and FeS$_2$ peaks detected in the energy 162 eV in S 2p$_{3/2}$ peak. considered that the FeS formed well during the nitriding process.

Fig. 7 shows main elements could act as the properties nitriding

the depth profiles of such as C, N, S that important roles to form of plasma sulfur process and they were

investigated by EPMA. In the plasma sulfur nitride sample, the S element diffused up to around 8 μm and the N element diffused up to 18 μm in the samples. The diffusion layers of S and N elements were almost the same as the thickness of FeS layer and that of the compound layer that were measured by optical microscope that is shown in Fig. 8. The distribution of C element was the higher at the top surface above the area where the highest S element was detected but it decreased very rapidly to the very small amount near to zero. Also below 20 μm from the surface, the content of nitrogen must be high enough for the hardened layer but the amount of C and that of S were very small and considered to be zero.

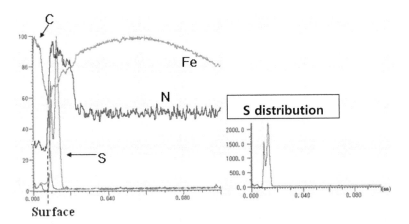

Figure 7. The depth profiles of the elements detected by EPMA

Fig. 8 is the optical microscopes showing cross sectional images of (a) M bushing and (b) S bushing. As previously mentioned, M bushing prepared by two steps of salt bath nitriding and MoS₂ coating process and S bushing was process by plasma nitriding and plasma sulfur nitrocarburizing in one chamber with different gas source. As a result of a nitriding process, both of the specimens had the compound layer with the thickness over 20 μm. At the top of the compound layer, M bushing had a MoS₂ layer and S bushing had a FeS layer with the thickness around 6~8 μm.

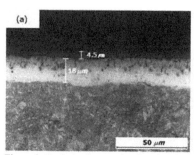 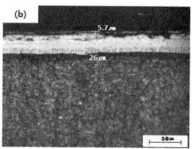

Figure 8. Optical micrographs showing the cross sectional images of the specimens; (a) M bushing and (b) S bushing

The Vickers microhardness profile of both specimens was measured to find out the thickness of the hardened layer and the data were shown in Fig. 9. The surface hardness of M bushing was over 900 Hv and this was higher than that of S bushing, about 800 Hv. This is because the MoS₂ coating layer is harder than FeS layer. But during the salt bath nitriding, the compound layer could be formed fast and easily but the diffusion of the nitrogen could not be possible. So in the M bushing, the hardness was decreased rapidly from 900 Hv at the surface region to 400 Hv corresponding to the hardness of quenching steels. So, it had only 250 μm for the hardened layer. During the plasma nitrocarburizing in S bushing, it was introduced the high nitrogen atmosphere by a subsidiary cathode process and this resulted in the high nitrogen contents and a subsequent high harness layer with 600 Hv and 100-200 μm below the compound layer. So in the S bushing, the hardness was decreased slowly from 800 Hv at the surface region to 400 Hv at the quenched structure region. The S bushing had a hardened layer with 400 μm thickness.

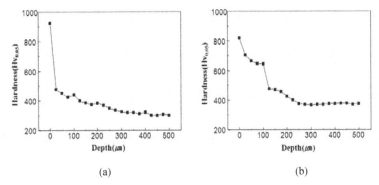

(a)   (b)

Figure 9. Microhardness profile of the specimens: (a) M bushing and (b) S bushing

In the B bushing that was prepared by carburizing and a Surf BT process, there was no compound layer in the B bushing. Only FeS layer with 10 μm thickness was formed at the surface and the surface hardness was 900 Hv. Also by a carburizing process, a hardened layer with the hardness over 550 Hv was formed with the thickness of 500-600 μm below the FeS layer. The microstructure and hardness profile of B bushing was presented in Fig. 10.

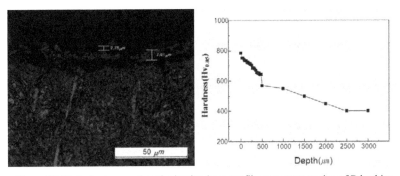

Figure 10. Optical micrograph and microhardness profile on a cross section of B bushing

3. 2. Endurance tests on the heat treated bushing

A set of endurance tests on the heat treated bushings were carried out by a rig tester in order to find out the life time of each bushing. By comparing the life time of S and B bushing with that of M bushing (commercialized one), it had decided on the possibility of the usage of S or B bushing as a commercial bushing. During the rig tests, the velocity was 1.24 m/min and the load was 750 kgf/cm^2. During the acceleration tests, adhesive wearing out would take place and this produce a progressive damage that cause both the torque and grease temperature to gradually increase. So, in this study if the friction coefficient would rapidly increase or it would be over 0.2 or if there were a sudden rise of grease temperature over 150 °C that the grease could be operated safely, it is considered that the bushing should be broke down in the test.

The data on the friction coefficient and the grease temperature during the acceleration tests are summarized in Fig. 11. Although there were some rising points in friction coefficient of S bushing, it was between 0.14 and 0.15 and almost similar with that of M bushing. But the occasional rising of friction coefficient of M bushing resulted in the different in the average friction coefficients of two bushing. That is, the average friction coefficient of M bushing and S bushing were 0.143 and 0.152, respectively. The difference in friction coefficient also affected on the grease temperature and the temperature of M bushing and S bushing were measured about 120 °C and 140 °C, respectively. The values were maintained until 4500 cycles in rig tests but there were sudden rising of friction coefficient and grease temperature between 4500 and 5000 cycles both in M busing and S bushing. Whereas, B bushing had a low friction coefficient as 0.15~0.16 at the early of the rig test but the friction coefficient was rapidly increased to higher than 0.2 at over 1500 cycles. Also, grease temperature of B bushing was over 150℃ at over 1500-2000 cycles. B bushing did not have enough endurance properties as to be used in the field but S bushing had an enough properties for the commercial usage, compared with M bushing. The detailed data of the endurance tests were summarized in Table 3.

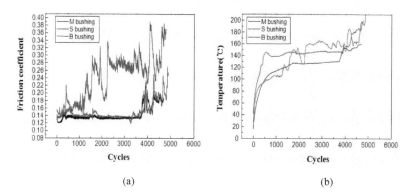

(a)   (b)

Figure 11. The endurance test data; (a) friction coefficient (b) temperature change

Table 3. The important values for the endurance test data

Factor	M bushing	S bushing	B bushing
Friction coefficient	0.143	0.152	0.16 (2000 cycles)
Surface temp. (℃)	123.4	146.7	>150
Life time (cycles)	4670	4880	<2000

	Shaft	Bush
M bushing		
S bushing		
H bushing		

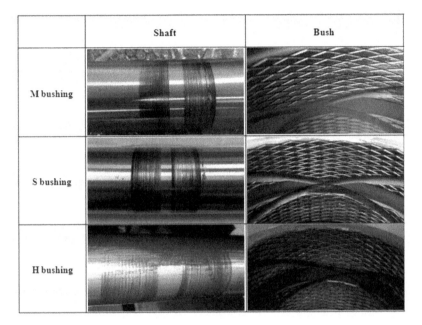

Figure 12. Surface images of the samples after wear tests

Fig. 12 shows the optical micrographs on the wearing surfaces in the shaft pins and bushing. The scar width in the pin for the S bushing was much shallower even compared with that of the M bushing. This must be related to the surface microstructure in S bushing. As shown in Fig. 3, the surface of S bushing consisted of very fine and porous granular structure. This fine and porous structure could be removed easily form the surface and attached the counterpart (the shaft pin) and it could protect the shaft during the endurance test [8]. So, comparing the wearing area in the counterpart of S busing even with that of M bushing, there was a large portion of undamaged metallic surface area in the pin of S bushing as shown in Figure 12. But the damage in the bushing, there was no difference between M bushing and S bushing. After the endurance tests, very severe wear in shaft and bushing was observed in the B bushing and this resulted in the lower service life of B bushing.

As the results of the endurance tests, we had the following conclusions on the properties of heat treated bushings.
- Friction coefficient: M bushing > S bushing > B bushing
- Grease Temperature: M Bushing > S bushing > B bushing
- Service time: S bushing > M bushing > B bushing

4. SUMMARY

In this study, it is found that the newly developed S bushing prepared by plasma sulfur nitrocarburizing could replace the commercialized M bushing with salt bath nitriding and MoS_2 coating. The important conclusions were made as follows.

*   The FeS and compound layer were formed on the surface of the SCM440 with thickness about 5 and 20μm. And the grain size of the FeS layer is less than 100 nm.
*   According to XRD analyses, a mixture of γ and ε nitride phase was observed in sulfur nitrocarburizing sample. But the formation of FeS layer was confirmed by XPS analyses.
*   According to EPMA analyses, the S element diffused up to around 6 μm and the C element was the highest than other elements on the top of the sample. But both of element decreased very rapidly to the zero concentration below 10 μm. The N element diffused though out the hardened layer below 20 μm and maintained high enough concentration for the hardness.
*   Although the S bushing had higher friction coefficient and resulting higher operating grease temperature, they had a longer expected service time that was determined on the point where a rapid increase of the friction coefficient occurred. This must be FeS layer on the top of the compound layer had enough endurance during the operation condition for the bushing. So it is considered that the S coating could be used for the busing instead of M coating.

5. REFERENCE

1. V.L.Kolmogorov, "Friction and wear model of heavy-loaded sliding pair Part II . Application to an unlubricated journal bearing" Wear, Vol.197, pp.9-16, 1996
2. Yoshikiyo Tanaka, "Development of new materials for special oil-impregnated bearings" KOMATSU Technical Report, Vol.49 No.152, pp.15-19, 2003
3. Ning Zhang, "Tribological behaviors of steel surfaces treated with ion sulphurization duplex processes" Surface and Coating Technology, Vol 203, Issues 20-21, pp.3173-3177, 2009
4. Monika Gierzynska-Dolna, "Effect of the surface layer in increasing the life of tools for plastic working" Journal of Mechanical Working Technology, Vol.6, Issues 2-3, pp.193-204, 1982
5. Fredrik Svahn, "The influence of surface roughness on friction and wear of machine element coatings" Wear, Vol.254, pp.1092-1098, 2003
6. N.M.Renevier, "Advantages of using self-lubricating, hard, wear-resistance MoS2-based coating" Surface and Coating Technology, Vol.142-144, pp.67-77, 2001
7. Naofumi Hiraoka, "Wear life mechanism of journal bearing with bonded MoS2 film lubricants in air and vacuum" Wear, Vol.249, pp.1014-1020, 2002
8. K. I. Moon, Y.K. Kim, K.S. Lee, "The formation of nanostructure compound layer during sulfur plasma nitriding and its mechanical properties" Ceramic Eng and Sci Proceedings, Vol 30, No 7, pp.99-109 (2010)
9. www.hef-group.com
10. B. Edenhofer, "Physical and metallurgical aspect of ion nitriding" Heat Treat Met., Vol. **2**, pp. 59-67 (1974).
11. M.B. Karamis, Tribological behavior of plasma nitrided 722M24 material under dry sliding condition, Wear, Vol. **147**, 385-99 (1991).
12. Da-Ming Zhuang, "Microstructure and tribological properties of sulphide coating produced by ion sulphuration" Wear, Vol. **225-229**, pp. 799-805 (1999)
13. Ning Zahang, "Effect of the substrate state on the microstructure and tribological properties of sulphide layer on 1045 steel" Applied Surface Science, Vol. **161**, pp. 170-177 (2000)

MICROSTRUCTURAL CHARACTERISATION OF POROUS TiO$_2$ CERAMIC COATINGS
FABRICATED BY PLASMA ELECTROLYTIC OXIDATION OF Ti

Po-Jen Chu, Aleksey Yerokhin[*], Allan Matthews
Department of Materials Science and Engineering, University of Sheffield
Sheffield, S1 3JD, U.K

Ju-Liang He
Department of Materials Science and Engineering, Feng Chia University
Taichung, 40724, Taiwan, R.O.C.

ABSTRACT
This study aims at providing a thorough examination of compositional and microstructural heterogeneities in the porous oxide ceramic surface layers formed on Ti by plasma electrolytic oxidation (PEO). The PEO-titania layers of up to 5μm thick were produced using 0.02...0.04M NaH$_2$PO$_4$ electrolyte solutions in the voltage range of 450...500V DC. Advanced methods of X-ray diffractometry (XRD), scanning electron microscopy (SEM) and field-emission transmission electron microscopy (FE-TEM) were employed to observe the surface layer morphology and characterise its crystal structure across the whole coating thickness. As revealed by XRD analysis, the surface layers consist of both anatase and rutile evenly distributed across the layer thickness. Detailed TEM studies showed that a continuous amorphous layer exists at the top of the PEO coating, the layer comprises mainly TiO$_2$, with some phosphorus incorporated from the electrolyte. Underneath, there is a porous crystalline layer with uniformly distributed nano-scale pores (<50nm), anatase nanocrystallites (<100nm) and submicrometer-scale (0.1 to 1μm) rutile crystallites. Large micrometre size pores (about 1...3μm in diameter) surrounded by nanocrystalline anatase are found to exist at the bottom of the porous crystalline layer, adjacent to the thin interfacial barrier layer. Correlations between thermal-physical properties of the coating material, heat dissipation conditions and microstructural evolution in the surface layer during coating formation are discussed.

INTRODUCTION
Plasma electrolytic oxidation (PEO) is an efficient method for production of functional ceramic surface layers on non-ferrous metals, such as aluminium,[1; 2] magnesium,[3] zirconium[4] and titanium[5] alloys. During PEO process the electrochemical reaction of anodic oxidation on the metal surface is assisted by micro-discharge events to promote formation of thick, hard and well-adhered oxide ceramic surface layers with specific morphologies and phase compositions, exhibiting superior protective performance and other useful electro-physical and chemical properties.[6]

Titanium currently attracts significant attention due to high specific strength, corrosion-resistance and biocompatibility and application of PEO treatment allows tailoring surface composition, crystallographic structure and morphology to attain a variety of functions that cannot be provided by the substrate metal alone. Versatility of PEO coatings on Ti together with simple processing set-up and low capital costs triggered many attempts of applying PEO to titanium for different purposes, including tribological,[7] gas-sensor,[8] biomedical,[9] dielectric,[10] photocatalytic[11] and photovoltaic applications.[12]

It has been commonly observed that the microstructure of the PEO layer strongly influences its properties such as mechanical,[13] photocatalytic behaviour,[11] biological performance[14] as well as many

---

[*]- Corresponding author: Tel.: +44 (0) 114 222 5510; E-mail address: A.Yerokhin@sheffield.ac.uk (A. Yerokhin)

other functional properties. Many of these are originated from and rely upon the porous nature of the PEO layer however a little attention is paid to the compositional and structural heterogeneities across the coating, particularly for the TiO$_2$/Ti system. This paper aims to reveal the through-thickness microstructure of porous ceramic PEO titania coatings on Ti by using several advanced material characterisation techniques.

EXPERIMENTAL

Grade II (99.99% pure) titanium plates (of dimension 25mm×20mm×0.5mm) were used as substrates. Prior to the coating, the samples were polished by #800 abrasive paper, cleaned with detergent, water rinsed, acetone flushed and air dried. The PEO treatment conditions and corresponding specimen designations are presented in Table I. The electrolyte bath was made up with aqueous solution of sodium dihydrogen phosphate dihydrate (NaH$_2$PO$_4$·2H$_2$O) of desired concentration and 0.5M NaOH added to adjust its pH value to 9.8. The samples immersed into the electrolyte were potentiostatically polarised by a DC power supply and a stainless steel plate was used as the cathode. The treatment was performed for 60min, during which the electrolyte temperature was maintained below 40°C by cooling system.

Table I.   PEO treatment conditions employed to form porous ceramic surface layers on Ti

Specimen designation	A	B	C
Electrolyte concentration (M)	0.02	0.04	0.02
Discharge voltage (V)	500	500	450
Discharge current (A)	0.48	1.2	0.2
Ultimate bath temperature(°C)	17	40	14
Phases identified by GA-XRD	Anatase+ Rutile	Anatase + Rutile	Anatase

A JEOL JSM 6400 scanning electron microscope (SEM) was used to observe surface morphology of the samples. The crystal structure was characterised using a Siemens D5000 X-ray diffractometer (CuK$_\alpha$ radiation), within 2θ scan range of 20° to 90°, with 0.02° step size. Diffraction patterns were taken at glancing angles (GA) ranging from 1° to 6°, to enable depth-resolved structural characterisation of PEO layers.

Detailed field-emission transmission electron microscopy (FE-TEM) analysis was carried out on a specific sample of interest (specimen B). The sample was manually polished and underwent a standard procedure of focused ion beam milling (FIB) using a FEI Nova-600 instrument. The FIB technique was selected to facilitate the preparation of TEM samples from porous and brittle ceramic materials. For the TEM study, a Tecnai™ G2 F-20 FE-TEM operated at an accelerating voltage of 200kV was used. A bright field TEM image was initially examined over the entire PEO layer to pin-point specific areas for further detailed microstructural characterisation using selected area diffraction (SAD) technique. High-resolution images were used to measure the d-spacings of the phase structures previously detected by SAD. Furthermore, high annular diffraction dark field (HADDF) analysis operated at scanning transmission electron microscope (STEM) was used to provide elemental distribution information over the entire PEO layer.

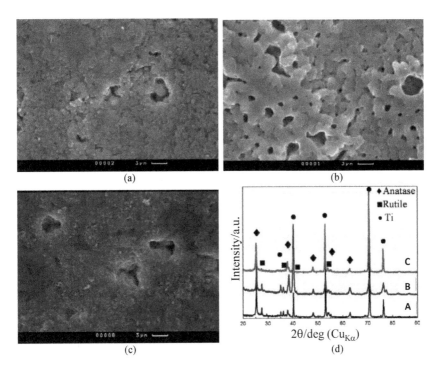

Figure 1. SEM surface morphology (a-c) and corresponding XRD patterns (d) of PEO coatings on samples: (a) A (500V; 0.02M), (b) B (500V; 0.04M); (c) C (450V; 0.02M).

RESULTS AND DISCUSSION

Crystal Structure and Surface Morphology of the PEO Layer

Surface morphologies and XRD patterns of the PEO treated specimens designated as A, B and C in Table I are shown in Figure 1. Samples A and B were obtained at 500V with a bath concentration 0.02 and 0.04M respectively. A higher voltage provides a higher power input which leads to higher electrolyte temperature and higher discharge currents for these two samples. It appears that both of them contain a mixture anatase and rutile phases in the PEO layer. This agrees well with reports that the increased voltage is associated with a larger discharge power favours the high-temperature rutile phase formation.[15] The SEM image of the PEO layer on sample B (Figure 1(b)) shows the most porous surface morphology among the three studied samples. It was therefore decided to use it for further GA-XRD analysis and TEM characterisation work to study local phase and pore distributions in the PEO layer, in more detail. Comparison between the samples A and C PEO-treated at 500 and 450V respectively shows that significant difference in discharge currents (0.48 and 0.2A, Table I) leads to corresponding difference in the surface layer phase composition. The anatase TiO$_2$ formation requires much lower activation energy than its rutile polymorph.[16; 17] Higher discharge currents or longer treatment durations thermodynamically favour the stable rutile phase. This agrees well with the results revealed by Lie, et al.[14]

Among the three samples shown in Figure 1, the surface morphology of sample B presents comparatively large, micrometre-scale voids. This is may be caused by a higher power provided to the growing PEO layer, with more relatively large micro-discharge events occurring, wherein more electrons are released to develop higher instantaneous currents, and consequently larger crater-like pores are formed in the centres of breakdown events. Additionally, small nanometre-scale voids uniformly distributed in the ceramic layer can be found, formed probably by thermal decomposition of hydrated oxides being heated up by the discharge and/or by releasing of dissolved gaseous products from the melt when the discharge is cooled down. Spatial distributions of the large and the small voids are addressed in subsequent TEM analysis.

Figure 2(a) shows GA-XRD diffraction patterns at different glancing angles for the sample B, from which a depth-resolved structure of the PEO layer composed of anatase and rutile can be determined. From high to low glancing angle, the intensities of the peaks ascribed to the rutile phase appear to reduce faster than those ascribed to the anatase, an implication of the anatase phase being more abundant in the outer region of the PEO layer. However, a more accurate estimate provided by the ratio of $I_{A(101)}/I_{R(110)}$ which was calculated as a function of the penetration depth in Figure 2(b), clearly shows that the $I_{A(101)}/I_{R(110)}$ ratio does not change significantly with the glancing angle. This allows suggesting that both anatase and rutile are distributed randomly across the PEO layer and the misleading intuition in Figure 2(a) shall be ascribed to the very strong $I_{A(101)}$ diffraction peak (even for the powder material, as follows from JCPDS No 89-4921).

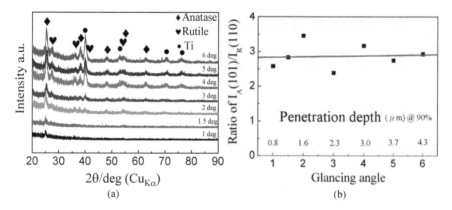

Figure 2. (a) GA-XRD patterns in different grazing angles of the sample B and (b) calculated $I_A(101)/I_R(110)$ ratio for each pattern shown in (a).

A low magnification cross-sectional bright field (BF) image and corresponding HADDF image of the ceramic surface layer on sample B are shown in Figure 3 (a) and (b), respectively. The HADDF

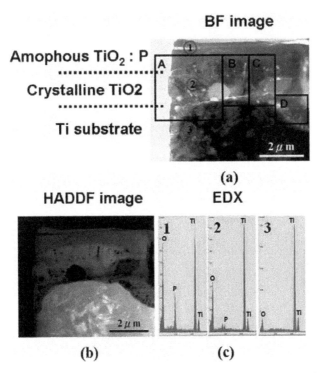

Figure 3. (a) Low magnification cross-sectional bright field image, (b) the corresponding HADDF image of the PEO layer in sample B and (c) EDX compositional point analysis of the PEO layer.

image is useful to identify the elemental distribution. The BF image shows clearly that an amorphous TiO$_2$ layer (denoted B1 in Figure 4) is present over the top of the PEO layer. This amorphous TiO$_2$ layer, however, cannot be resolved by GA-XRD and contains a considerable amount of phosphorus, as shown by EDX analysis in Figure 3(c). A speculation can be made that this is a result of unsatisfactory cleaning of the PEO treated sample prior to TEM sample preparation, which leaves phosphorus-containing electrolyte residues over the sample surface. In fact however, the sample underwent many cleaning cycles after the PEO treatment and is unlikely to retain any absorbed electrolyte. Our explanation is that during the prolonged PEO treatment under the potentiostatic conditions, the oxide layer growth achieves a steady state wherein a relative mobility of species in the plasma discharge becomes important. The adsorbed phosphate anions involved into discharge are likely to be decomposed and may yield ionic phosphorus which would tend to migrate outwards, following the electric field in the similar way as titanium cations. Compared to titanium however, the size of phosphorus cations is smaller and the charge is higher which makes them more mobile. With time, this leads to the enrichment of the outer coating layers with phosphorus and this, together with rapid cooling by the electrolyte, promotes amorphisation of the coating material in that region. This finding is in agreement with the work by Matykina, et al,[18] where a mixed (NaPO$_3$)$_6$/CaHPO$_4$ electrolyte was employed for PEO treatment of titanium under a DC voltage of 340V.

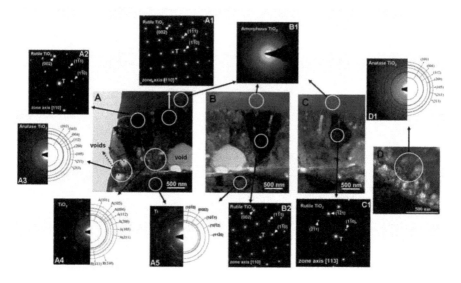

Figure 4. SAD patterns obtained from zone A, B, C and D bright field images for the PEO layer as segmented.

Elemental Composition and Pore Distribution in the PEO Layer

Another observation in Figure 3(a) is that underneath the amorphous layer comes a crystalline form of TiO$_2$ where a vast amount of pores of different sizes are present. The pores appear dark in the HADDF image shown in Figure 3(b). It is also interesting to find that small (<50nm in diameter) and large pores tend to be distributed in the outer and the inner coating regions respectively. This observation supports our understanding of the effects that physical and chemical processes associated with discharge phenomena have on the porosity formation in PEO coatings developed during discussion of Figure 1. Since titanium exhibits passive behaviour in the alkaline phosphate environment, a thin barrier layer of titanium dioxide grows slowly at the metal/electrolyte interface up to the potentials of breakdown.[19] Microdischarge events occurring as a result of breakdown lead to the formation of additional oxide material which is deposited on the top of the barrier layer. Therefore, large pores and a thin barrier layer at the interface with substrate can be observed in the PEO coating. It is also noticeable that a very large (relative to the large closed pore located in the centre) pore exists at the right side of the image (half shown) providing an open channel to the top of the PEO layer. The observation of such pore, of several microns in diameter, is consistent with the SEM observation of surface morphology in Figure

1(b) where the largest pore (in the middle of the micrograph) is about 3μm in diameter. In terms of porosity, it can be summarised that the PEO layer features closed individual pores or patulous channels of different size penetrating throughout the coating up to the barrier layer. It seems to follow the micro-discharge event which is determined by the dielectric breakdown mechanism.

Detailed Crystal Structural Characterisation of the PEO Layer

The analysis is performed in segmented zones A, B, C and D shown in Figures 3(a) and 4.

Zone A. In zone A, an SAD pattern corresponding to the bottom of the bright field (BF) image verifies presence of α-Ti phase corresponding to the metal substrate (shown by the ring pattern in Figure 4, A5). The SAD pattern obtained from the top area of zone A shows the amorphous coating material (ring pattern in Figure 4, B1). Two SAD patterns corresponding to the crystalline region at a distance away from the large pores reveal rutile TiO$_2$ structure with high crystallinity (ring pattern in Figure 4, A1 and A2). The grain size is estimated to be in the submicron range. In the regions close to and around the periphery of the large pores, very fine grains of anatase structure are found and identified by SAD (ring pattern Fig. 4 A4). They are estimated to be <100nm in diameter. This would be logical, because the metastable anatase TiO$_2$ is easier to form under non-equilibrium conditions. This is also shown by Ma, et al,[20] who used a high-energy laser beam to partially transform the stable rutile into the anatase phase. In their case, extremely high heating and cooling rates (about 10^{4}K/s) provided by the laser were apparently responsible for the non-equilibrium phase transformation. The crystallites at a distance away from the large pore have enough time to grow a stable rutile phase.

Zones B and C. Separate crystallites found in the middle region of the PEO layer and a small distance away from large pores were analysed. Both SAD analyses performed on zones B and C clearly demonstrate that the rutile phase is evenly distributed in the middle of the PEO layer (ring patterns in Figure 4, B2 and C1). The estimated grain size of these rutile crystallites is similar to those found in the zone A.

Zone D. The SAD analysis was performed at the bottom of the particular large pore. It is interesting to notice that the phase identified is anatase with very fine grains (shown by the ring pattern Figure 4 D1). This particular pore seems to be generated by a micro-discharge event at later coating growth stage, with the crystal structure of surrounding material resembling that around the periphery in zone A. It seemed that very high cooling rates associated with large micro-discharge events favour the formation of metastable anatase structures, which is consistent with the coating formation mechanisms discussed above.

High-resolution images obtained across the PEO layer are shown in Figure 5. Features of amorphous TiO$_2$ can be found in area A. The transition from amorphous titania into crystalline rutile is clearly seen in area B, while anatase lattice in area C around the pore periphery is identified. These verify the findings in the SAD analysis.

Microtructural Evolution of the PEO Layer

From the above TEM analysis, a phase zone diagram of the studied coating region of can be developed as shown in Figure 6. The coating comprises a thin dense barrier layer and a porous part composed by the three main phases, namely anatase, rutile and amorphous Ti-P-O compound. The barrier layer is located at the interface with Ti substrate, whereas anatase phase is distributed in the middle region along with numerous small voids of 200 to 500nm in diameter. Several large rutile grains (1 to 2μm in diameter) are located in denser parts of the middle region which also features several large voids appear beside the anatase regions and rutile grains. At the top, a continuous amorphous layer (several hundred microns in thickness) is located, covering the coating surface.

Figure 5. High-resolution images sampled from the top to the bottom areas of the PEO layer.

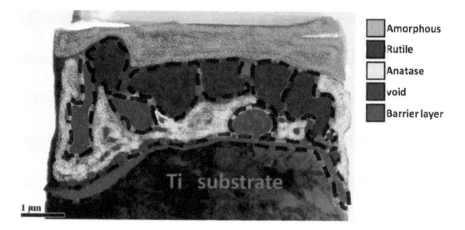

Figure 6. A phase zone diagram of the coating.

Based on the diagram in Figure 6, significant heterogeneities in the crystal structure and spatial distribution of porosity across the coating can be revealed. This is considered to be mainly affected by heat dissipation during the growth of the porous ceramic layer with low thermal conductivity, as proposed in Figure 7. It is generally accepted that the PEO coating provides minimum contribution to the heat flow dissipation, as indicated by the arrows in Figure 7. For thin coatings, the heat provided by discharge is easily exchanged through the electrolyte. Therefore amorphous compounds are preferably formed, especially in the outer regions of the PEO coating subjected to rapid cooling. Once the amorphous layer is formed it provides a thermal barrier between the rest of the coating and the electrolyte which facilitates heat accumulation within the coating and promotes 'hot enough' microdischarges to form crystalline rutile grains, while leaving a nanometre-scale porosity surrounded by fine anatase grains in the reminder of the middle part of the coating. During further coating growth, the heat exchange in it is redistributed leading to more thermal energy accumulated in the middle part of the coating which causes yet more powerful discharge events producing larger open porosity. This increases specific surface area of the PEO lyer and allows more electrolyte permeation into the coating, promoting formation of anatase nano-crystallites due to substantially high cooling rates in the vicinity of the large pores.

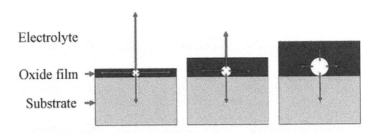

Figure 7. Schematic of the heat dissipation flow associated with discharge events occurring at different growth stages.

CONCLUSIONS

Detailed characterisation has been carried out throughout the thickness of a porous TiO$_2$ ceramic surface layer formed by potentiostatic DC PEO treatment of Ti in an alkaline-phosphate electrolyte. The distribution of the composition and the crystal structure across the coating is discussed and the microtructural evolution is proposed as follows.

A continuous amorphous TiO$_2$ layer was found over the top of the PEO coating with phosphorus incorporated from the electrolyte. The middle coating region is formed by uniformly distributed nano-scale pores (<50nm), anatase nanocrystallites (<100nm) and sub-micrometre scale rutile crystallites. Large micrometre size pores surrounded by nanocrystalline anatase TiO$_2$are found to exist in the inner part of the middle region which is separated from the metal substrate by a thin interfacial barrier layer.

The microstructural evolution in the surface layer may be associated with dielectric and thermal-physical properties of the porous oxide material as well as with changes in the conditions of heat dissipation following micro-discharge events occurring at different stages of the PEO treatment.

ACKNOWLEDGEMENT

Financial support provided for AY by the UK EPSRC is acknowledged with thanks. The authors are also grateful to Mr. Y.-F. Ko and Mr. V. Lee (Materials Analysis Technology, Inc. Hsinchu, Taiwan) for help in providing FIB and TEM facilities as well as to Mr. W.-L. Wang (National Chiao Tung University, Taiwan) for help with sample preparation and characterisation.

REFERENCES

[1] A. L. Yerokhin, X. Nie, A. Leyland, A. Matthews, and S. J. Dowey, Plasma electrolysis for surface engineering, *Surf. Coat. Technol.*, **122**, 73-93 (1999).

[2] A. L. Yerokhin, A. A. Voevodin, V. V. Lyubimov, J. Zabinski, and M. Donley, Plasma electrolytic fabrication of oxide ceramic surface layers for tribotechnical purposes on aluminium alloys, *Surf. Coat. Technol.*, **110**, 140-46 (1998).

[3] S. V. Gnedenkov, O. A. Khrisanfova, A. G. Zavidnaya, S. L. Sinebryukhov, V. S. Egorkin, M. V. Nistratova, A. Yerokhin, and A. Matthews, PEO coatings obtained on an Mg-Mn type alloy under unipolar and bipolar modes in silicate-containing electrolytes, *Surf. Coat. Technol.*, **204**, 2316-22 (2010).

[4] M. Klapkiv, N. Povstyana, and H. Nykyforchyn, Production of conversion oxide-ceramic coatings on zirconium and titanium alloys, *Mater. Sci.*, **42**, 277-86 (2006).

[5] A. L. Yerokhin, X. Nie, A. Leyland, and A. Matthews, Characterisation of oxide films produced by plasma electrolytic oxidation of a Ti-6Al-4V alloy, *Surf. Coat. Technol.*, **130**, 195-206 (2000).

[6] A. L. Yerokhin, Plasma electrolysis - Preface, *Surf. Coat. Technol.*, **199**, 119-20 (2005).

[7] L. Ceschini, E. Lanoni, C. Martini, D. Prandstraller, and G. Sambogna, Comparison of dry sliding friction and wear of Ti6Al4V alloy treated by plasma electrolytic oxidation and PVD coating, *Wear*, **264**, 86-95 (2008).

[8] Y. Han, S. H. Hong, and K. W. Xu, Porous nanocrystalline titania films by plasma electrolytic oxidation, *Surf. Coat. Technol.*, **154**, 314-18 (2002).

[9] H.-T. Chen, C.-H. Hsiao, H.-Y. Long, C.-J. Chung, C.-H. Tang, K.-C. Chen, and J.-L. He, Micro-arc oxidation of [beta]-titanium alloy: Structural characterization and osteoblast compatibility, *Surf. Coat. Technol.*, **204**, 1126-31 (2009).

[10] J. H. Peng, B. Han, W. F. Li, J. Du, P. Guo, and D. H. Han, Study on the microstructural evolution of BaTiO3 on titanium substrate during MAO, *Mater. Lett.*, **62**, 1801-04 (2008).

[11] J. F. Li, L. Wan, and J. Y. Feng, Study on the preparation of titania films for photocatalytic application by micro-arc oxidation,   pp. 2449-55.

[12] W.-C. L. Shu-Yuan Wu, Keh-Chang Chen, Ju-Liang He, Study on the preparation of nano-flaky anatase titania layer and their photovoltaic application,   (2008).

[13] P. Huang, F. Wang, K. W. Xu, and Y. Han, Mechanical properties of titania prepared by plasma electrolytic oxidation at different voltages,   pp. 5168-71 in 5th Asian/European International Conference on Plasma Surface Engineering.

[14] L. H. Li, Y. M. Kong, H. W. Kim, Y. W. Kim, H. E. Kim, S. J. Heo, and J. Y. Koak, Improved biological performance of Ti implants due to surface modification by micro-arc oxidation, *Biomaterials*, **25**, 2867-75 (2004).

[15]Y. Han, S. H. Hong, and K. W. Xu, Structure and in vitro bioactivity of titania-based films by micro-arc oxidation, *Surf. Coat. Technol.,* **168**, 249-58 (2003).

[16]U. Bach, D. Lupo, P. Comte, J. E. Moser, F. Weissortel, J. Salbeck, H. Spreitzer, and M. Gratzel, Solid-state dye-sensitized mesoporous TiO2 solar cells with high photon-to-electron conversion efficiencies, *Nature,* **395**, 583-85 (1998).

[17]B. Oregan and M. Gratzel, A Low-Cost, High-Efficiency Solar-Cell Based on Dye-Sensitized Colloidal Tio2 Films, *Nature,* **353**, 737-40 (1991).

[18]E. Matykina, R. Arrabal, P. Skeldon, and G. E. Thompson, Transmission electron microscopy of coatings formed by plasma electrolytic oxidation of titanium, *Acta Biomaterialia,* **5**, 1356-66 (2009).

[19]Y. T. Sul, C. B. Johansson, S. Petronis, A. Krozer, Y. Jeong, A. Wennerberg, and T. Albrektsson, Characteristics of the surface oxides on turned and electrochemically oxidized pure titanium implants up to dielectric breakdown: the oxide thickness, micropore configurations, surface roughness, crystal structure and chemical composition, *Biomaterials,* **23**, 491-501 (2002).

[20]H. L. Ma, J. Y. Yang, Y. Dai, Y. B. Zhang, B. Lu, and G. H. Ma, Raman study of phase transformation of TiO2 rutile single crystal irradiated by infrared feratosecond laser, *Appl. Surf. Sci.,* **253**, 7497-500 (2007).

Author Index